Coastal
Meteorology

Coastal
Meteorology

S. A. Hsu

COASTAL STUDIES INSTITUTE
SCHOOL OF GEOSCIENCE
LOUISIANA STATE UNIVERSITY
BATON ROUGE, LOUISIANA

Academic Press, Inc.
Harcourt Brace Jovanovich, Publishers
San Diego   New York   Berkeley   Boston
London   Sydney   Tokyo   Toronto

ACADEMIC PRESS, INC.
1250 Sixth Avenue
San Diego, California 92101

United Kingdom Edition published by
ACADEMIC PRESS INC. (LONDON) LTD.
24-28 Oval Road, London NW1 7DX

Library of Congress Cataloging-in-Publication Data

Hsu, S. A. (Shih-Ang)
  Coastal meteorology.

  Bibliography: p.
  Includes index.
  1. Meteorology, Maritime. 2. Coasts. I. Title.
QC994.H78 1988   551.6914′6   88-3288
ISBN 0-12-357955-4 (alk. paper)

PRINTED IN THE UNITED STATES OF AMERICA
88 89 90 91   9 8 7 6 5 4 3 2 1

Contents

v

Chapter 5   Synoptic Meteorology

Chapter 6   Atmospheric Boundary Layers and Air–Sea Interaction

Chapter 7   Air–Sea–Land Interaction

Chapter 8   Engineering Meteorology

Preface

Coastal meteorology is an integral part of the total-system approach to understanding coastal environments. Those who work in the coastal zone, such as oceanographers, geologists, biologists, and engineers, have found a need for a reference or textbook on the subject.

Since 1970 I have taught a course titled "Coastal and Marine Meteorology" in the Department of Marine Sciences at Louisiana State University, for graduate students and upper-level undergraduates in science and engineering. I taught similar courses as Visiting Professor in the Department of Coastal and Oceanographic Engineering at the University of Florida in Gainesville in the summer of 1985 and as a Visiting Scientist at the Earth Resources Lab of NASA in the summer of 1986. Throughout those years I searched in vain for a suitable textbook for such a one-semester course. At the urging of many other professors, coastal researchers, and former students, I have organized my lecture notes into this book. My goal is to provide information for those students who are not necessarily majoring in meteorology or atmospheric sciences but who nonetheless have need of such knowledge. Scientists, engineers, coastal planners, and the like whose curricula might not include meteorology, particularly of the coastal zone, may wish to consider this book as a useful resource for familiarizing themselves with meteorological information. This book is designed to be largely self-contained [e.g., the mathematical derivation of the Gaussian (or normal) distribution for atmospheric dispersion estimates is given], so that anyone who has had first-year college physics and calculus should be able to follow it.

I am indebted to a number of scientists and engineers who made this book possible. I wish especially to thank Professor Amos Eddy and to acknowledge the influence of the late Professor Bernhard Haurwitz, who introduced me to the detailed physics of thermally driven coastal circulation (the land–sea breeze system) and the many meteorological subdisciplines while I was a graduate student at the University of Texas at Austin (from 1965 through 1969).

Many of the results incorporated in this book were based on field experiments funded since 1969 by, among others, the National Science Foundation,

Office of Naval Research, Naval Environmental Prediction Research Facility, and National Oceanic and Atmospheric Administration through the Coastal Studies Institute (CSI), Louisiana State University. All of this support and the excellent interdisciplinary research group and facilities at CSI are greatly appreciated.

The following publishers are acknowledged for permission to use illustrations: Academic Press, American Chemical Society, American Geophysical Union, American Meteorological Society, D. Reidel Publishing Company, Elsevier Science Publishers B. V., John Wiley and Sons, Louisiana State University Press, Louisiana State University School of Geoscience, University of Puerto Rico, U. S. Government Agencies, U. S. Naval Institute, and World Meteorological Organization.

The cover was designed by Mrs. Wanda Huh.

Shih-Ang (S.A.) Hsu

Chapter 1 | Introduction

The name meteorology comes from the Greek *meteoros*, meaning lofty, and *logos*, meaning a formal treatment of a subject (Wallace and Hobbs, 1977). Thus, meteorology is the study of atmospheric phenomena and their spatial and temporal behavior.

1.1 Composition of the Atmosphere

The earth's atmosphere near its surface consists of a mixture of gases and solid or liquid particles. Table 1.1 indicates that approximately 99% of the air is made up of nitrogen and oxygen, but in most meteorological processes these

Table 1.1

Composition of the Atmosphere

Constituent	Percent by volume	Molecular weight
Major gases		
Nitrogen (N_2)	78.08	28.02
Oxygen (O_2)	20.95	32.00
Argon (A)	0.93	39.94
Variable gases		
Water vapor (H_2O)	0–4%	18.02
Ozone (O_3)	0–12 ppm[a]	48.00
Carbon dioxide (CO_2)	325 ppm	44.01
Other gases		
Neon (Ne), helium (He). krypton (Kr), hydrogen (H), xenon (Xe), methane (CH_4), and nitrous oxide (N_2O)		

[a] Parts per million.

1

elements are quite passive. On the other hand, water vapor plays a dominant role in thermodynamic processes because it can change phases, such as the formation of clouds, snow, and ice from water vapor. Carbon dioxide and ozone are also important because of their role in radiative processes. No other gas is important meteorologically.

1.2   The Physical Foundation

To analyze and understand the behavior of the atmosphere, basic laws of physics are applied. We are concerned with radiation, thermodynamics, and dynamics.

Radiation is the study of energy transfer or heat exchange transmitted by electromagnetic waves. Thermodynamics deals with the initial and final equilibrium states produced by energy processes or transformation. Dynamics, or hydrodynamics, is the study of the motion of fluids in relationship to the forces acting upon them. Topics on radiation, thermodynamics, and dynamics are discussed in order in the next three chapters.

1.3   Scales of Atmospheric Motion

Atmospheric phenomena occur on a very wide range of time and space scales. Figure 1.1 depicts these variabilities. The characteristic sizes of these

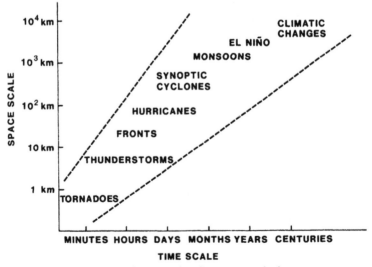

Fig. 1.1. Time and space scales of some atmospheric systems.

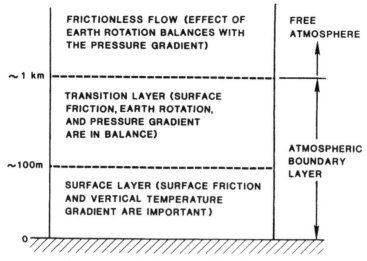

Fig. 1.2. Atmospheric layers over flat and open regions.

phenomena vary from a few centimeters to several thousand kilometers. Time scales may vary from a few seconds, as with turbulence, to several weeks, as with the general circulation around the globe.

Air movement over a homogeneous region is shown in Fig. 1.2.

1.4 The Air–Sea–Land Boundary

In the coastal zone, where air, sea, and land meet, air movement is drastically changed from the simple picture of Fig. 1.2. An example is shown in Fig. 1.3. One of the most important changes is the development of an internal boundary layer over land when the wind blows from sea to land.

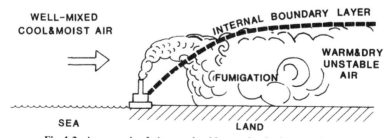

Fig. 1.3. An example of air–sea–land interaction in the coastal zone.

1.5   Scope of Coastal Meteorology

Since meteorology is the study dealing with the phenomena of the atmosphere, coastal meteorology may be defined as that part of meteorology that deals mainly with the study of atmospheric phenomena occurring in the coastal zone. This description includes the influence of atmosphere on coastal waters and the influence of the sea surface on atmospheric phenomena, i.e., air–sea interaction.

The behavior of the atmosphere can be analyzed and understood in terms of basic laws and concepts of physics. The three fields of physics that are most applicable to the atmosphere are radiation, thermodynamics, and hydrodynamics. Chapters 2–4 will deal with these subjects.

1.6   Interdisciplinary Approach

Since meteorology is an integral part of the total-systems approach to understanding coastal environments, emphasis is given not only to studies related to meteorology itself but also to its application to other disciplines such as oceanography, geology, biology, and engineering.

Chapter 5 introduces synoptic meteorology to these interdisciplinary realms. Chapter 6 concerns the subject of air–sea interaction. Topics in air–sea–land interaction are discussed in Chapter 7. For practical applications, engineering meteorology for the coastal zone is covered in Chapter 8.

Chapter 2 | Radiation

2.1 Some Physics of Radiation

Radiation is a form of energy transfer or heat exchange transmitted by electromagnetic waves. Radiation may be characterized by its wavelength, which is measured in terms of the micrometer ($1 \ \mu m = 10^{-4}$ cm). The spectrum of radiation ranges from very short waves such as X-rays ($< 10^{-3} \ \mu m$), through the visible (4×10^{-1} to $8 \times 10^{-1} \ \mu m$), to infrared ($4 \times 10^{2} \ \mu m$), microwaves (10^{2} to $10^{7} \ \mu m$), and radio waves ($> 10^{7} \ \mu m$).

Since all bodies emit radiation if their absolute temperatures are above zero degree Kelvin (0 K), a hypothetical "body," the so-called "blackbody," is employed in radiation studies. If a body at a given temperature emits the maximum possible amount of radiation per unit of its surface area per unit time, then it is called a blackbody. A blackbody is also a body that absorbs all of the electromagnetic radiation striking upon it. This definition does not imply that the object must be black in color; for example, snow is an excellent blackbody in the infrared part of the spectrum.

According to Planck's law, the amount of radiation emitted by a blackbody is determined by its temperature T (see, e.g., Wallace and Hobbs, 1977):

$$E_\lambda = C_1 \lambda^{-5} [e^{(C_2/\lambda T)} - 1]^{-1}$$
$$= C_1 \lambda^{-5} e^{-C_2/\lambda T} \tag{2.1a}$$

where E_λ is the monochromatic irradiation in watts per square meter per micrometer, λ is the wavelength, $C_1 = 3.74 \times 10^{-16}$ W m^{-2}, and $C_2 = 1.44 \times 10^{-2}$ m K.

The total blackbody radiation over all wavelengths at a given temperature can be obtained by interpreting Eq. (2.1a) such that

$$E = \int_0^\infty E_\lambda \, d\lambda$$

$$= \sigma T^4 \tag{2.1b}$$

where E is the flux of radiation in langley (ly) units ($= 1$ cal/cm^2) per unit of time such as a minute; T is the absolute temperature in degrees Kelvin (K); and σ is the Stefan–Boltzmann constant, which is equal to 0.813×10^{-10} cal/cm^2 min K^4, or 5.67×10^{-8} W m^{-2} K^{-4}. For explanation of units, see Appendix A. Equation (2.1b) is called the Stefan–Boltzmann law.

It can be shown that the wavelength of peak energy for a blackbody at temperature T is given by (see, e.g., Byers, 1974)

$$\left(\frac{\partial E}{\partial \lambda}\right)_T = 0 \quad \text{or} \quad \lambda_m = 2897/T \tag{2.2}$$

where λ_m is measured in micrometers and T is in kelvins. Equation (2.2) is called Wien's displacement law. Through Eq. (2.2), it is possible to estimate the temperature of a radiation source from knowledge of its emission spectrum.

2.2 Solar and Terrestrial Radiation

Experimental evidence indicates that Eqs. (2.1b) and (2.2) are good approximations for the sun and the earth. Since the temperature of the sun is about 6000 K, and according to Eq. (2.2) the peak wavelength is around 0.47 μm. On the other hand, if one uses the temperature of the earth, say 27°C ($\simeq 300$ K), the peak wavelength is around 10 μm. Thus, radiation from the sun is also called shortwave radiation as compared to that from the earth, that is, longwave radiation.

Because of the difference in solar and terrestrial temperature, there is a tremendous disparity in the amounts of energy emitted; that is, from Eq. (2.1b),

$$\frac{E_{sun}}{E_{earth}} = \frac{\sigma T_{sun}^4}{\sigma T_{earth}^4}$$

$$= \left(\frac{6000}{300}\right)^4$$

$$= 160,000$$

In other words, the energy emitted per unit area per unit time by the sun is 160,000 times as great as that radiated by the earth. Since the total surface area of the sun is 10,000 times greater than that of the earth, the total energy emitted by the sun is 1.6 billion times as great as that emitted by the earth. However, since the earth receives only a small part of the sun's total radiation, the amount of solar energy absorbed by the earth–atmosphere system is exactly equal to that emitted by the earth–atmosphere system.

According to Sellers (1965), the sun provides about 99.97% of the heat energy required for the physical processes taking place in the earth–atmosphere system. Each minute it radiates approximately 56×10^{26} cal of energy. In terms of the energy per unit area incident on a spherical shell with a radius of 1.5×10^{13} cm (the mean distance of the earth from the sun) and concentric with the sun, this energy is equal to

$$S = \frac{56 \times 10^{26} \quad \text{cal min}^{-1}}{4\pi (1.5 \times 10^{13} \quad \text{cm})^2}$$

$$= 2.0 \quad \text{ly min}^{-1}$$

where S is the solar constant.

Latitude, season, time of day, and cloud cover are variables that will result in a reduction in energy available to a horizontal surface. Three processes contribute to the depletion of solar energy by the atmosphere: absorption by ozone, water vapor, CO_2, and clouds; scattering by air molecules, water droplets, and very small particles; and reflection. Since Rayleigh scattering is inversely proportional to the fourth power of the wavelength, the sky appears blue because there is more scattering of blue light than there is of other colors in the longer wavelengths. If the entire wavelength is being affected equally, large solid particles in the atmosphere reflect rather than scatter light. The blue color of the sky therefore turns to white with increasing pollution. Note that the addition of scattering and reflection equals the diffused solar radiation, which is the total shortwave energy received in the shade.

Another term commonly used in meteorology is albedo (A), which is the ratio of the flux of solar radiation diffused by a surface Q_R to the flux incident upon it, that is, $A = Q_R/Q_T$. Therefore, the solar radiation available for energy transformation at the earth's surface is $(1 - A)Q_T$.

2.3 Radiation and Heat Balance

All gains and losses of radiative energy at the earth's surface must balance. The net radiation may be obtained by the following equation (see, e.g.,

Munn, 1966):

$$Q_N = Q_T - Q_R + Q_{L\downarrow} - Q_{L\uparrow} \tag{2.3}$$

where Q_T is the shortwave radiation from the sun and sky, assumed by convention to be positive. It is absent at night. The term Q_R is the shortwave radiation reflected from the sea surface. It is absent at night. The term $Q_{L\downarrow}$ is the longwave radiation received by the surface from the atmosphere, and $Q_{L\uparrow}$ is the longwave radiation emitted by the surface.

The heat balance for a given system such as the air–sea interface implies that the sum of all heat sources is equal to the heat sinks. That is (see, e.g., Colon, 1963),

$$Q_r = Q_e + Q_s + Q_v + Q_t \tag{2.4}$$

where Q_r represents the net gain of heat by radiative processes, Q_e is the heat lost to the atmosphere by evaporation, Q_s is the heat lost to the atmosphere by conduction, Q_v is the divergence of heat transported by ocean currents, and Q_t is the time rate of change of the heat content of the water, i.e., the storage term.

Examples of the heat balance in various coastal environments are discussed in detail in the following subsections.

2.3.1   Heat Balance of a Semienclosed Sea

The Caribbean Sea is used in this example. To compute the heat balance for the Caribbean Sea, Eq. (2.4) can be written in the form (see, e.g., Colon, 1963)

$$Q_a = Q_e + Q_s = Q_r - Q_v - Q_t \tag{2.5}$$

where

$$Q_e = LE = \rho_{air} C_D L(q_{sea} - q_{air})U \tag{2.6}$$

$$Q_s = \rho_{air} C_p C_D(T_{sea} - T_{air})U \tag{2.7}$$

$$Q_v = MC(\bar{T}_y - \bar{T}_e) \tag{2.8}$$

and

$$Q_t = C \int_\alpha \frac{\partial T_{sea}}{\partial t} \rho_{sea} \, d\alpha \tag{2.9}$$

where L stands for the heat of vaporization, E for evaporation rate, C_D for the drag coefficient (see Chapter 6), C_p and ρ_{air} for specific heat and density of air, ρ_{sea} for density of water, q_{sea} for saturation specific humidity of the water, q_{air} for specific humidity of the air (see Chapter 3), U for wind speed (normally at 10 m above the surface), T_{sea} and T_{air} for temperatures of the sea surface and air, respectively, M for the integrated water-mass transport across the oceanic

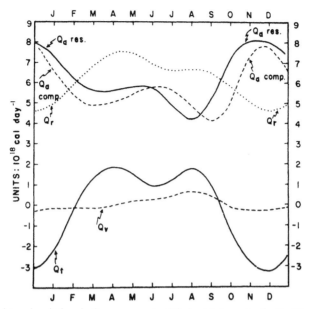

Fig. 2.1. Annual variations in the components of the heat balance for the Caribbean Sea. Solid curve on bottom, rate of change in heat content of the water due to seasonal temperature change (Q_t); dashed curve on bottom, net divergence of heat transport by Caribbean current (Q_v); dotted curve on top, net radiation absorbed by Caribbean Sea (Q_r); solid curve on top, net transfer to atmosphere (Q_a), obtained as residual in the heat balance; dashed curve on top, net transfer to atmosphere, computed with turbulent transfer formulas for section north of the Canal Zone. [After Colon (1963).]

current boundaries (assumed to be the same in the Antilles and in the Yucatan Channel), $\bar{T}_y$ and $\bar{T}_e$ for mean temperatures over the current in the Yucatan Channel and in the Antilles (east side), C for the specific heat of the water, and α for the volume. Note that both C and ρ_{sea} may be taken as unity.

Annual variations in the components of the heat balance for the Caribbean Sea are shown in Fig. 2.1. They are explained as follows:

Q_r: Net absorption of radiation. The total radiation absorbed by the Caribbean Sea ranges from 4.6 units (in 10^{18} cal day^{-1}) in December to about 7.5 units in April–May, with an annual average of about 6.3 units.

Q_t: Seasonal energy storage in the water. The rate of change of heat content is zero in late September and late February. There is a very rapid decrease in late fall and early winter, but the increase during the spring and early summer is more gradual. The maximum rate of cooling amounts to -3.1 units in December, and the maximum warming rate is $+1.8$ units in April and August.

Q_v: Divergence of heat transport. According to Eq. (2.8), the heat transported by the current into and out of the Caribbean is small because the difference in surface water temperatures between the Yucatan Channel and the Antilles (east side) is small.

Q_a: Heat transfer to atmosphere, that is, the net loss of heat to the atmosphere by evaporation [Eq. (2.6)] and conduction [Eq. (2.7)].

The results show maximum values of total heat transfer for the Caribbean Sea of about 8.0 units in November and December and minimum of 4.2 units in August. Therefore, the seasonal variations seem to be determined mostly by the variation of Q_t. Note that because the difference between T_{sea} and T_{air} is small, Q_s, the sensible heat, is only about 10% of Q_e, the latent heat.

Because the values of Q_e range from 7.35 units in November to 3.89 units in August, one gets from Eq. (2.6) that the evaporation rates are 0.58 cm day^{-1} in November and 0.31 cm day^{-1} in August. The annual evaporation rate is 0.44 cm day^{-1}, or 161 cm per year. This rate is less than half that obtained from semiarid areas such as the Gulf of Aqaba, where the annual evaporation rate is equivalent to 1 cm day^{-1} or 365 cm per year (see Assaf and Kessler, 1976).

2.3.2   Energy Balance of a Swamp

A swamp is a low, wet area covered by trees or tall bushes. Mangrove swamps occur widely in tropical climates. In this example, the mangrove swamp on Grand Cayman is used.

The net radiation absorbed by the canopy of the mangrove swamp (R) can be calculated by Eq. (2.3). The procedure has been outlined in Hsu *et al.* (1972a). The result is shown in Fig. 2.2, where $R = R_N$. The net radiation near the floor of the mangrove swamp (R_0) may be calculated by [cf. Eq. (2.4)]

$$R_0 = G + H + LE \qquad (2.10)$$

where G is the soil heat flux, H is the sensible heat flux, and LE is the contribution of latent heat of evapotranspiration.

In Fig. 2.2 two features may be seen:

1. Near the floor of the mangrove swamp on Grand Cayman the soil heat (G) and sensible heat (H) fluxes may be neglected inasmuch as they are much smaller in amount than the contribution of latent heat of evapotranspiration (LE). Thus

$$R_0 = LE \qquad \text{or} \qquad E = R_0/L \qquad (2.11)$$

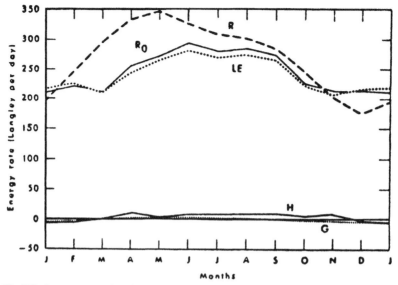

Fig. 2.2. Average annual variation of the main components of the energy balance of the mangrove swamp on Grand Cayman Island. In the figure R represents the net radiation absorbed by the canopy of the mangrove swamp, R_0 is that absorbed by the floor, LE is the contribution of the latent heat of evapotranspiration, H is the sensible heat flux, and G is the soil heat flux (Hsu et al., 1972a).

The quantity E may also be regarded as E_0, the potential evapotranspiration, which represents the evapotranspiration that would occur from a large, fully wetted surface, that is, one with an unlimited water supply. Then Eq. (2.11) may be written as

$$E_0 = R_0/L \qquad (2.12)$$

Equation (2.12) is in conformity with the results obtained by others (see, e.g., Budyko, 1956).

2. From the difference between the net radiation absorbed by the canopy of the mangrove swamp (R) and that by the floor (R_0), as shown in Fig. 2.2, it may be concluded that photosynthesis and heat storage within trees may be important, in particular during the growing season, as suggested by Munn (1966).

2.3.3   Heat Balance on a Beach

Hsu (1980) reported on measurements of evaporation (E), net radiation (R), and soil heat flux (G) made in July on a windward beach of a tropical island (Barbados, West Indies) under prevailing trade winds. Sensible heat

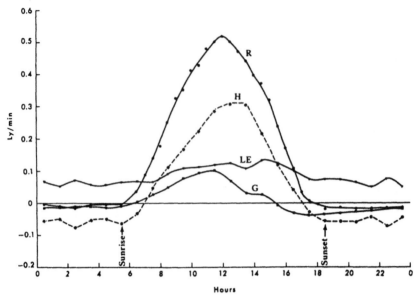

Fig. 2.3. Average hourly transfer of heat on the beach at Bath, Barbados, from July 25 to 30, 1973. (See text for explanation.)

flux (H) was estimated as a residual of the heat balance equation. The results, shown in Fig. 2.3, reveal that at midday approximately 60% of R (0.5 ly min^{-1}) was transferred to the atmosphere by H ($= 0.3$ ly min^{-1}), 20% by G ($= 0.1$ ly min^{-1}), and 20% by the latent heat flux ($LE = 0.1$ ly min^{-1}). Patterns of H and G reasonably followed R throughout the day, except that the peak of G was offset toward the morning and that of H toward the afternoon. At night, R, H, and G were negative and small. Because the diurnal variation of LE was much smaller than that of H, a constant value for the Bowen ratio, H/LE, was not warranted for the beach environment. The average daily net absorbed radiation, as computed from the radiation balance, is found to be in fair agreement with the measurement.

2.4   Remote Sensing from Space

Remote sensing from space, such as from meteorological satellites, is based on the fact that all bodies having a temperature greater than 0 K ($-273°$C) emit some radiation. This phenomenon is employed in infrared photography and is utilized in temperature determination by infrared techniques.

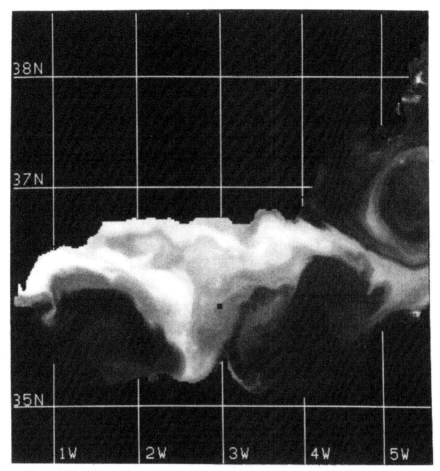

Fig. 2.4. The Alboran Sea gyre on October 11, 1982, as seen by the NOAA-6 advanced very high resolution radiometer (AVHRR) at approximately 1500 hr LT (local time) (except at 0300 LT on October 14). The data have been atmospherically corrected and registered to Mercator projection to allow continuity in temporal and spatial comparison. (Courtesy of Paul E. La Violette of U.S. Naval Ocean Research and Development Activity.)

Figure 2.4 shows as an example the difference in water temperature in the Alboran Sea, between Spain and Morocco, on October 10 and 11, 1982. Figure 2.5 illustrates the sea-surface temperature and winds at 300 m on October 11, 1982, for the area. Information on sea-surface temperatures are routinely available in many parts of the world, such as the Gulf Stream system.

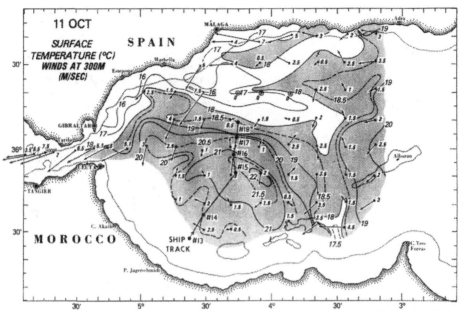

Fig. 2.5. Sea-surface temperature distribution in the Alboran Sea as derived from aircraft radiometer temperature taken from a height of 300 m. Winds at 300 m as measured from the same aircraft are superimposed. [From Hsu (1984a).]

Chapter 3 | Atmospheric Thermodynamics

3.1 The Equation of State

Charles's law states that at constant pressure P the specific volume α is directly proportional to the absolute temperature T, that is,

$$\frac{\alpha_T}{T} = \frac{\alpha_0}{T_0} \qquad (3.1)$$

where T_0 is 273 K and α_T and α_0 are specific volumes at temperature T and T_0, respectively.

Boyle's law states that pressure and volume are inversely proportional to each other, that is,

$$P\alpha = \text{constant}$$
$$= f(T) \qquad (3.2)$$

From Eqs. (3.1) and (3.2) the equation of state of an ideal gas can be derived:

$$P\alpha = RT$$
$$= \frac{R^*}{m} T \qquad (3.3)$$

where R is the specific gas constant for the gas being considered. The specific gas constant for dry air is 2.87×10^6 erg g^{-1} K^{-1}. The term R^* is the universal gas constant, equal to 8.3144×10^7 erg mol^{-1} K^{-1}. The parameter m is the molecular weight of a sample of gas under consideration.

3.2 Some Principles of Thermodynamics

3.2.1 Heat

When two fluids of different temperature are mixed, the temperature difference diminishes. The final temperature will be the average of the two initial temperatures. If T_{warm} is the initial temperature of the warm body and T_{cold} is the initial temperature of the cold body, the final or mean temperature at thermal equilibrium T_{mean} will be $T_{warm} > T_{mean} > T_{cold}$. Thus

$$C_1(T_{mean} - T_{warm}) + C_2(T_{mean} - T_{cold}) = 0$$

where the constants C_1 and C_2 are the heat capacities of the two fluids.

The amount of heat ΔH lost by the warm body is

$$-\Delta H = C_1(T_{mean} - T_{warm})$$

and the amount of heat gained by the cold body is equal in magnitude and given by

$$\Delta H = C_2(T_{mean} - T_{cold})$$

Thus

$$C = \frac{\Delta H}{\Delta T} \simeq \frac{dH}{dT}$$

if ΔT becomes infinitesimal.

The specific heat capacity at constant volume for dry air is

$$C_v = \left(\frac{dh}{dT}\right)_\alpha = 0.171 \text{ cal g}^{-1} \, {}^\circ\text{C}^{-1}$$

and the specific heat capacity at constant pressure for dry air is

$$C_p = \left(\frac{dh}{dT}\right)_p = 0.240 \text{ cal g}^{-1} \, {}^\circ\text{C}^{-1}$$

Note that $C_v \, dT$ also represents the internal energy (see, e.g., Hess, 1959) dE or

$$dE = C_v \, dT \tag{3.4}$$

3.2.2 The First Law of Thermodynamics

When a force of magnitude F is applied to a mass, which then moves through a distance dX parallel to the force, the work done is $dW = F \, dX$.

Since $F = P\,dA$, where P is the pressure and dA is the surface element, we have

$$dW = F\,dX = P\,dA\,dX = P\,dV \tag{3.5}$$

where dV is the infinitesimal change in the volume in question.

The first law of thermodynamics states that

$$dH = dE + dW \tag{3.6}$$

where dH is the increment of heat added, dE is the internal energy, and dW is the work done.

For unit mass and from Eqs. (3.4) and (3.5), Eq. (3.6) becomes

$$dh = C_v\,dT + P\,d\alpha \tag{3.7}$$

From Eq. (3.3) we have

$$P\alpha = RT$$

or

$$P\,d\alpha + \alpha\,dP = R\,dT \tag{3.8}$$

Substituting Eq. (3.8) into Eq. (3.7), we get

$$dh = (C_v + R)\,dT - \alpha\,dP \tag{3.9}$$

For an isobaric process, that is, $dP = 0$, we have

$$C_P = \left(\frac{dh}{dT}\right)_P = C_v + R$$

or

$$C_P - C_v = R \tag{3.10}$$

Substituting Eq. (3.10) into Eq. (3.9), the results show

$$dh = C_p\,dT - \alpha\,dP \tag{3.11}$$

Equation (3.11) is commonly used in meteorology because dT and dP can be measured by thermometers and barometers.

3.2.3 Adiabatic Processes

If no heat is added to or taken away from a sample of gas, this thermodynamic process is called adiabatic. Under this condition, Eq. (3.11) becomes

$$dh = 0 = C_p\,dT - \alpha\,dP$$

but $\alpha = RT/P$ [cf. Eq. (3.3)]; thus

$$\frac{dT}{T} - \frac{R}{C_p}\frac{dP}{P} = 0$$

Since $R/C_p = \kappa = 0.286$ (for dry air), we have

$$\int_{T_1 = \theta}^{T_2 = T} \frac{dT}{T} = \kappa \int_{P_1 = 1000\,mb}^{P_2 = P} \frac{dP}{P} \tag{3.12}$$

where θ is the potential temperature, defined as the temperature at $P = 1000$ mb. One millibar (mb) $= 10^3$ dyn cm^{-2}. Note that near sea level $P \simeq 1000$ mb (1 atm $= 1013.25$ mb).

After integration, Eq. (3.12) becomes

$$\ln T - \ln \theta = \kappa(\ln p - \ln 1000)$$

or

$$\ln\frac{T}{\theta} = \kappa \ln\frac{P}{1000}$$

or

$$\frac{T}{\theta} = \left(\frac{P}{1000}\right)^\kappa \tag{3.13}$$

when a term such as $\ln T$ is called the natural logarithm of T. Note that during adiabatic processes θ is invariant.

3.3 Some Aspects of Moist Air

3.3.1 The Equation of State of Moist Air

Experiments have shown that dry air and water vapor separately satisfy the equation of state of an ideal gas with sufficient accuracy for our purposes. Since moist air is composed of dry air and water vapor, the mean molecular weight of moist air is given by (see, e.g., Hess, 1959)

$$\frac{1}{\bar{m}} = \frac{1}{M_d + M_v}\left(\frac{M_d}{m_d} + \frac{M_v}{m_v}\right)$$

or

$$\frac{1}{\bar{m}} = \frac{1}{m_d}\frac{M_d}{M_d + M_v}\left(1 + \frac{M_v/M_d}{m_v/m_d}\right) \tag{3.14}$$

where M is the mass of each constituent in a mixture, m is the molecular weight of that constituent, and subscripts d and v stand for dry and moist air, respectively. The ratio M_v/M_d is the mass of water vapor per unit mass of dry air. This nondimensional measure of the moisture content of air is called the mixing ratio.

The equation of state of water vapor is [cf. Eq. (3.3)]

$$e = \rho_v \frac{R^*}{m_v} T \tag{3.15}$$

where e is the vapor pressure and ρ_v is the corresponding density. Note that $\rho\alpha = 1$.

The equation of state of dry air is

$$(P - e) = \rho_d \frac{R^*}{m_d} T \tag{3.16}$$

From Eqs. (3.15) and (3.16) the mixing ratio M_v/M_d may be written as

$$w = \frac{M_v/V}{M_d/V} = \frac{\rho_v}{\rho_d} = \frac{m_v}{m_d}\left(\frac{e}{P - e}\right)$$

Since the molecular weight of the water vapor m_v is 18 g mol^{-1} and the mean molecular weight of the dry air M_d is approximately 29 g mol^{-1}, the ratio ε ($=m_v/m_d$) is thus approximately 0.62. Therefore, the mixing ratio is

$$w = 0.62 \frac{e}{P - e} \tag{3.17}$$

Now, if we divide both numerator and denominator on the right-hand side of Eq. (3.14) by M_d, we have

$$\frac{1}{\bar{m}} = \left(\frac{1}{m_d}\right)\frac{1}{1 + M_v/M_d}\left(1 + \frac{M_v/M_d}{m_v/m_d}\right) \tag{3.18}$$

Substituting $M_v/M_d = w$ and $m_v/m_d = \varepsilon$ into Eq. (3.18), we get

$$\frac{1}{\bar{m}} = \left(\frac{1}{m_d}\right)\frac{1}{1 + w}\left(1 + \frac{w}{\varepsilon}\right)$$

Therefore, the equation of state of moist air is

$$P\alpha = \frac{R^*}{\bar{m}} T = \frac{R^*}{m_d}\left(\frac{1 + w/\varepsilon}{1 + w}\right) T$$

Note that this equation is the same as the equation of state for dry air except for the correction term shown in parentheses.

If we define a fictitious temperature that satisfies the equation of state,

$$T^* \equiv \left(\frac{1 + w/\varepsilon}{1 + w}\right) T \qquad \text{or} \qquad T^* = (1 + 0.61w)T$$

where $\varepsilon = 0.62$ is used, then [cf. Eq. (3.3)]

$$P\alpha = \frac{R^*}{m_d} T^* = R_d T^*$$

or

$$P = \rho R_d T^* \tag{3.19}$$

This fictitious temperature is called virtual temperature, which is obviously the temperature the dry air would have if its pressure and specific volume were equal to those of a given sample of moist air.

Since w never exceeds 40 g of water vapor per kilogram of dry air (see, e.g., Hess, 1959), that is, $w \le 0.04$,

$$T^* = \frac{1 + 1.609 \times 0.04}{1 + 0.04} T$$

$$= \frac{1.06}{1.04} T$$

Thus for practical use $T^* \simeq T$.

3.3.2   Moisture Parameters

Saturation vapor pressure e_s is defined as the vapor pressure of a system, at a given temperature, that has attained saturation, that is, the vapor pressure reaches its upper limit. The following equation may be used for computing e_s (see List, 1951):

$$e_s = 6.1078 \times 10^{[7.5T/(237.3 + T)]} \tag{3.20}$$

where e_s is the saturation vapor pressure over water in millibars at temperature $T(°C)$. When the wet-bulb temperature T' is substituted for T, the saturation vapor pressure e_{sw} at T' is obtained.

The vapor pressure e can be obtained from

$$e = e_{sw} - 0.00066(1 + 0.00115T')P(T - T') \tag{3.21}$$

where e is the existing vapor pressure (mb) in the air at pressure p (mb), temperature T (°C), and wet-bulb temperature T' (°C).

Dew point, or dew-point temperature T_{dew}, is defined as the temperature to which a given parcel of air must be cooled at constant pressure and constant water-vapor content in order for saturation to occur. It may be computed from

$$T_{dew} = \frac{237.3 \log_{10}(e/6.1078)}{7.5 - \log_{10}(e/6.1078)} \tag{3.22}$$

where T_{dew} is the dew point in degrees Celsius. Note that Eqs. (3.20) and (3.22) are the same if T is substituted for T_{dew} and e_s for e.

All the equations above are applicable to wet- and dry-bulb psychrometry when the wet bulb is not frozen. The formula for e is valid when the wet bulb is frozen if e_{si} is substituted for e_{sw}:

$$e_{si} = 6.1078 \times 10^{[9.321 T_i'/(261.24 + T_i')]} \tag{3.23}$$

where e_{si} is the vapor pressure over ice in millibars at temperature T_i', the temperature of the frozen wet-bulb thermometer in degrees Celsius. Then

$$T_{D,i} = \frac{261.24 \log_{10}(e/6.1078)}{9.321 - \log_{10}(e/6.1078)} \tag{3.24}$$

where $T_{D,i}$ is the dew-point temperature as determined from the wet- and dry-bulb psychrometer when the wet bulb is frozen.

Specific humidity q is defined as the ratio of the mass of water vapor to the mass of moist air containing the vapor. Thus

$$q = \frac{M_v}{M_v + M_d} = \frac{\rho_v}{\rho_v + \rho_d} = \varepsilon \frac{e}{P} \tag{3.25}$$

where $\varepsilon = 0.62$.

Mixing ratio w is the ratio of the mass of water vapor present to the mass of dry air containing the vapor. Thus, from Eq. (3.17),

$$w = \frac{\rho_v}{\rho_d} = \varepsilon \frac{e}{p - e} \tag{3.26}$$

The saturation mixing ratio is given by

$$w_s = \varepsilon \frac{e_s}{p - e_s} \tag{3.27}$$

where e_s can be obtained from Eq. (3.20).

Relative humidity RH is the ratio of the actual mixing ratio of a sample of air at a given temperature and pressure to the saturation ratio of the air at that temperature and pressure:

$$RH = \frac{w}{w_s} \tag{3.28}$$

where RH is a percentage and w and w_s can be obtained from Eqs. (3.26) and (3.27), respectively.

Wet-bulb temperature T_{wet} is defined as the lowest temperature to which air can be cooled by evaporating water into it at constant pressure, where all the heat of vaporization comes from the air.

3.4   Atmospheric Stability

3.4.1   The Hydrostatic Equation

The hydrostatic equilibrium states that balance exists between the force of gravity g and the vertical pressure force $\alpha\,dP/dZ$, that is,

$$-\alpha\frac{dP}{dZ} = g$$

or

$$\frac{dP}{dZ} = -\rho g \tag{3.29}$$

Equation (3.24) is called the hydrostatic equation. The term $\alpha\,dP/dZ$ is obtained first by the consideration of vertical pressure gradient, that is, the rate of change of pressure P with height Z. Since pressure is a force per unit horizontal area, dP/dZ is a force per unit volume. To convert to force per unit mass, dP/dZ should be multiplied by the specific volume, that is, α.

The minus sign is necessary because dP/dZ is negative and g is a positive number.

3.4.2   The Hypsometric Equation

The hydrostatic equation can be used to compute a relationship between height and pressure. Since ρ is not commonly measured, we use the equation of state for moist air, Eq. (3.19). Eliminating ρ from Eqs. (3.19) and (3.29) shows that

$$\frac{dP}{dZ} = -\frac{Pg}{R_d T^*}$$

or

$$\int_{P_1}^{P_2} \frac{dP}{P} = -\frac{g}{R_d}\int_{Z_1}^{Z_2} \frac{dZ}{T^*}$$

$$\ln P_2 - \ln P_1 = -\frac{g}{R_d \overline{T}^*}(Z_2 - Z_1)$$

or

$$Z_2 - Z_1 = \frac{R_d \bar{T}^*}{g} \ln \frac{P_1}{P_2} \tag{3.30}$$

where $\bar{T}^*$ is the mean virtual temperature between the lower level, 1, and the upper level, 2. Note that $Z_2 > Z_1$ and $P_1 > P_2$, because the pressure decreases upward while elevation increases with height.

Equation (3.30) is called the hypsometric equation. Substituting proper values of R_d and g as given previously, this equation normally becomes

$$\Delta H = 14.636(T_1^* + T_2^*) \ln \frac{P_1}{P_2} \tag{3.31}$$

where ΔH is the thickness in meters of a stratum of air. The subscripts 1 and 2 denote values at the base and top, respectively, of the stratum. It is assumed that temperature varies linearly through the stratum. Then

$$H_n = \sum_{i=1}^{n} (\Delta H)_i$$

where H_n is the height at the top of the nth stratum.

3.4.3 The Dry Adiabatic Lapse Rate

Under adiabatic conditions, the potential temperature θ is a constant. Rearranging Eq. (3.13) gives

$$\ln T - \ln \theta = \kappa(\ln P - \ln 1000) \tag{3.32}$$

Now, logarithmic differentiation of Eq. (3.32) with respect to height z and holding θ constant yields

$$\frac{1}{T}\frac{dT}{dz} = \frac{\kappa}{P}\frac{dp}{dz}$$

Since $dp/dz = -\rho g$ [see Eq. (3.29)] and $\kappa = R/C_p$, we have

$$\frac{1}{T}\frac{dT}{dz} = \frac{\kappa}{P}(-\rho g)$$

or

$$\frac{dT}{dz} = -\frac{g}{C_p}\left(\frac{R\rho T}{P}\right)$$

Since $P = \rho R T$, the equation of state, we get

$$\frac{dT}{dz} = -\frac{g}{C_p} = -\frac{9.76}{1000}\frac{°C}{m}$$

or

$$\gamma_d \equiv -\frac{dT}{dz} = \frac{1}{100}\frac{°C}{m} \tag{3.33}$$

Equation (3.33) is called the dry adiabatic lapse rate, that is, the rate of decrease of temperature with height of a parcel of dry air lifted adiabatically through an atmosphere in hydrostatic equilibrium. This thermodynamic process [Eq. (3.33)] is very useful in meteorology.

3.4.4   The Moist Adiabatic Lapse Rate

The moist or saturation or pseudo-adiabatic lapse rate is defined as the rate of decrease of temperature with height of an air parcel lifted in a saturation adiabatic process through an atmosphere in hydrostatic equilibrium. Owing to the release of latent heat, this lapse rate is less than the dry adiabatic lapse rate (Huschke, 1959).

From Eq. (3.11) we have

$$dh = C_p\,dT - \alpha\,dP$$

But $dP = -\rho g\,dz$ [cf. Eq. (3.29)] and $\alpha\rho = 1$. The above equation becomes

$$dh = C_p\,dT + g\,dz \tag{3.34}$$

Now, if the saturation mixing ratio of the air with respect to water is w_s, the quantity of heat dh released into (or absorbed from) a unit mass of dry air because of condensation (or evaporation) of liquid water is $-Lw_s$, where L is the latent heat of condensation (Wallace and Hobbs, 1977). Therefore

$$-L\,dw_s = C_p\,dT + g\,dz \tag{3.35}$$

Dividing both sides of Eq. (3.35) by $C_p\,dz$ and rearranging terms, we get

$$\frac{dT}{dz} = -\frac{L}{C_p}\frac{dw_s}{dz} - \frac{g}{C_p}$$

or

$$\left(1 + \frac{L}{C_p}\frac{dw_s}{dT}\right)\frac{dT}{dz} = -\gamma_d$$

Thus the saturation adiabatic lapse rate γ_s is defined by

$$\gamma_s \equiv -\frac{dT}{dz} = \frac{\gamma_d}{1 + (L/C_p)(dw_s/dT)} \tag{3.36}$$

where γ_d, the dry adiabatic lapse rate, from Eq. (3.33), is equal to g/C_p.

Since $P \gg e_s$ [cf. Eq. (3.27)] and from Eq. (3.20), dw_s/dT is always positive; thus $\gamma_s < \gamma_d$ [from Eq. (3.36)]. Actual values of γ_s range from about 4°C/km near the ground in warm, humid air masses where dw_s/dT is very large, to typical values of 6–7°C/km in the middle troposphere (Wallace and Hobbs, 1977).

3.4.5 Stability Criteria

The stability criteria may be expressed in terms of lapse rate of potential temperature $d\theta/dZ$ since

$$\theta = T\left(\frac{1000}{P}\right)^{R/C_P}$$

or

$$\ln \theta = \ln T + \frac{R}{C_P}(\ln 1000 - \ln P)$$

Differentiating the above equation with respect to z gives

$$\frac{1}{\theta}\frac{d\theta}{dz} = \frac{1}{T}\frac{dT}{dz} - \frac{R}{C_P}\left(\frac{1}{P}\right)\frac{dP}{dz}$$

but since $dP/dz = -\rho g$ [from Eq. (3.29)] and $P = \rho RT$ [Eq. (3.3)], we have

$$\frac{1}{\theta}\frac{d\theta}{dz} = \frac{1}{T}\left(\frac{dT}{dz} + \frac{g}{C_P}\right)$$

Since the actual lapse rate of the air $\gamma = -dT/dz$ and the dry adiabatic lapse rate is $g/C_P = \gamma_d$,

$$\frac{T}{\theta}\frac{d\theta}{dz} = (\gamma_d - \gamma) \qquad (3.37)$$

Therefore, for unsaturated air, stability criteria are:

Lapse rate			Stability
$\gamma > \gamma_d$	or	$\dfrac{d\theta}{dz} < 0$	Unstable
$\gamma = \gamma_d$	or	$\dfrac{d\theta}{dz} = 0$	Neutral
$\gamma < \gamma_d$	or	$\dfrac{d\theta}{dz} > 0$	Stable

It can also be shown (see, e.g., Haltiner and Martin, 1957) that for saturated air

$$\gamma > \gamma_s \qquad \text{Unstable}$$

$$\gamma = \gamma_s \qquad \text{Neutral}$$

$$\gamma < \gamma_s \qquad \text{Stable}$$

where γ_s is the saturated adiabatic lapse rate [cf. Eq. (3.36)].

3.4.6 Thermodynamic Diagram and Its Applications

A thermodynamic diagram or adiabatic chart is a graphical display of the lines representing major atmospheric processes such as isobaric, isothermal, dry adiabatic, and saturation adiabatic. One of the most commonly employed charts is the so-called skew $T-\log P$ diagram, with temperature as abscissa and the logarithm of pressure (i.e., $\log P$), increasing downward, as ordinate. Dry adiabats (a line of constant potential temperature) and saturation (or pseudo) adiabats (a line representing pseudo-adiabatic expansion of an air parcel) are curved (see Fig. 3.1). Note that in this diagram the isotherms are rotated 45 degrees clockwise (thus the skewed T) to produce greater separation of isotherms and adiabats.

The thermodynamic chart is a very useful tool for studying meteorological processes, particularly for determining the lifting condensation level, the convective condensation level, and the level of free convection. All three levels are delineated in Fig. 3.1.

The lifting condensation level (LCL) is the level to which unsaturated air would have to be raised in a dry-adiabatic expansion to produce condensation. At the LCL the mixing ratio becomes equal to the saturation mixing ratio. Graphically, it is the level at which the dry adiabat through the initial pressure and temperature of the parcel intersects the saturation mixing ratio line, whose value of w_s is equal to the actual initial mixing ratio of the parcel (Hess, 1959).

The convective condensation level (CCL) on a thermodynamic diagram is the point of intersection of a sounding curve (representing the vertical distribution of temperature in an atmospheric column) with the saturation mixing ratio line corresponding to the average mixing ratio in the surface layer (i.e., approximately the lowest 500 m). The dry adiabat through this point determines, approximately, the lowest temperature to which the surface air must be heated before a parcel can rise dry adiabatically to its lifting condensation level without ever being colder than the environment. This temperature, the convective temperature, is a useful parameter in forecasting the onset of convection (Huschke, 1959). The CCL and the convective temperature, $T_{\text{convective}}$, are shown in Fig. 3.1.

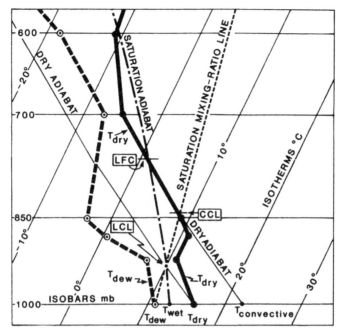

Fig. 3.1. The skew $T-\log P$ thermodynamic diagram (see text for explanation).

The level of free convection LFC is the level at which a parcel of air lifted dry adiabatically until saturated (i.e., at LCL), and saturation adiabatically thereafter would first become warmer than its surroundings in a conditionally unstable atmosphere. On a thermodynamic diagram the LFC is given by the point of intersection of the process curve, representing the process followed by the ascending parcel, and the sounding curve, representing the lapse rate of temperature in the environment. From the LFC to the point where the ascending parcel again becomes colder than its surroundings, the atmosphere is characterized by latent instability. Throughout this region the parcel will gain kinetic energy as it rises (Huschke, 1959). An example of LFC is shown in Fig. 3.1.

Another application of the thermodynamic chart is to determine the wet-bulb potential temperature θ_w, which is the temperature an air parcel would have if cooled from its initial state adiabatically to saturation, that is, at LCL, and thence brought to 1000 mb by a saturation adiabatic process.

Chapter 4 | Atmospheric Dynamics

4.1 The Continuity Equation

Consider a coordinate system with the x axis directed toward the east, y to the north, and z as the vertical. Their corresponding velocities are $dx/dt = U$, $dy/dt = V$, and $dz/dt = W$, where t represents time. Consider now some fluid property, such as density ρ, which is a function of x, y, z, and t. From calculus we know that

$$d\rho = dt\frac{\partial\rho}{\partial t} + dx\frac{\partial\rho}{\partial x} + dy\frac{\partial\rho}{\partial y} + dz\frac{\partial\rho}{\partial z} \tag{4.1}$$

where the partial differential has the usual meaning, that all other independent variables are being held as constants while the partial derivative is being evaluated.

Dividing Eq. (4.1) by dt, one gets

$$\frac{d\rho}{dt} = \frac{\partial\rho}{\partial t} + \frac{dx}{dt}\left(\frac{\partial\rho}{\partial x}\right) + \frac{dy}{dt}\left(\frac{\partial\rho}{\partial y}\right) + \frac{dz}{dt}\left(\frac{\partial\rho}{\partial z}\right)$$

or

$$\frac{d\rho}{dt} = \frac{\partial\rho}{\partial t} + u\frac{\partial\rho}{\partial x} + v\frac{\partial\rho}{\partial y} + w\frac{d\rho}{\partial z} \tag{4.2}$$

The derivative on the left-hand side, $d\rho/dt$, is called the individual change. The first term on the right, $\partial\rho/\partial t$, is the rate of change of the density with respect to time at a fixed point, that is, the local derivative or local change. Terms on the right, $u\,\partial\rho/\partial x$, etc., are advective changes caused by movement of air particles of differing density.

The equation of continuity is a hydrodynamical equation that expresses the principle of the conservation of mass in a fluid. It equates the increase in mass in a hypothetical fluid volume to the net flow of mass into the volume. There are several ways to derive the continuity equation. One of the most direct methods is employed here.

When the cylinder ends "move with the fluid" (i.e., Lagrangian), then the fluid velocity at the end point is

$$dX_A/dt = U_A \qquad dX_B/dt = U_B$$

From Fig. 4.1, the mass of the cylinder M is

$$M(t) = \int_{X_A(t)}^{X_B(t)} \rho(x, t)\, dx$$

Because of the conservation of mass,

$$dM/dt = 0$$

Applying Leibnitz's rule, we get

$$\frac{d}{dt} \int_{X_A(t)}^{X_B(t)} \rho\, dx = \int_{X_A}^{X_B} \frac{\partial \rho}{\partial t}\, dx + \rho_{XB} \frac{dX_B}{dt} - \rho_{XA} \frac{dX_A}{dt}$$

$$= \int_{X_A}^{X_B} \frac{\partial \rho}{\partial t}\, dx + \rho_B U_B - \rho_A U_A$$

$$= \int_{X_A}^{X_B} \frac{\partial \rho}{\partial t}\, dx + \int_{X_A}^{X_B} \frac{\partial \rho u}{\partial x}\, dx$$

$$= \int_{X_A}^{X_B} \left(\frac{\partial \rho}{\partial t} + \frac{\partial \rho u}{\partial x} \right) dx$$

$$= 0$$

Therefore,

$$\frac{\partial \rho}{\partial t} + \frac{\partial \rho u}{\partial X} = 0$$

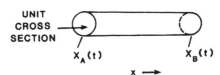

UNIT CROSS SECTION

$X_A(t)$ $X_B(t)$

$x \longrightarrow$

Fig. 4.1. A schematic for fluid motion in the x direction.

This is the equation of continuity in the x direction. For all three directions in x, y, and z we have

$$\frac{\partial \rho}{\partial t} + \frac{\partial \rho u}{\partial x} + \frac{\partial \rho v}{\partial y} + \frac{\partial \rho w}{\partial z} = 0$$

This equation may be expanded into

$$\frac{\partial \rho}{\partial t} + u\frac{\partial \rho}{\partial x} + v\frac{\partial \rho}{\partial y} + w\frac{\partial \rho}{\partial z} + \rho\frac{\partial u}{\partial x} + \rho\frac{\partial v}{\partial y} + \rho\frac{\partial w}{\partial z} = 0$$

From Eq. (4.2), we get

$$\frac{d\rho}{dt} + \rho\frac{\partial u}{\partial x} + \rho\frac{\partial u}{\partial y} + \rho\frac{\partial w}{\partial z} = 0$$

or

$$-\frac{1}{\rho}\left(\frac{d\rho}{dt}\right) = \frac{\partial v}{\partial x} + \frac{\partial v}{\partial y} + \frac{\partial w}{\partial z} \tag{4.3}$$

Equation (4.3) is the continuity equation, used extensively in meteorology. This equation requires that the individual rate of change of density be proportional to the three-dimensional velocity divergence.

A fluid in which an individual parcel experiences no change of density with time is said to be incompressible, that is, $d\rho/dt = 0$. In other words, if one assumes that the velocity divergence is much more variable than the density with respect to time, we have

$$\frac{\partial u}{\partial x} + \frac{\partial v}{\partial y} + \frac{\partial w}{\partial z} = 0$$

or

$$\frac{\partial u}{\partial x} + \frac{\partial v}{\partial y} = -\frac{\partial w}{\partial z} \tag{4.4}$$

which states that horizontal divergence must be compensated by vertical shrinking or vertical convergence. Conversely, horizontal convergence must be accompanied by vertical stretching or vertical divergence.

4.2   The Equations of Motion

The equations of motion are a set of hydrodynamical equations representing the application of Newton's second law of motion. They state that the rate of change of momentum of a body with time is equal to the vectoral sum of all forces acting upon the body and is in the same direction:

$$\frac{d}{dt}(m\mathbf{V}) = \sum \mathbf{F} \tag{4.5}$$

where m is the mass of the body, $\mathbf{V}$ is its velocity, t is time, $\mathbf{F}$ is force, $\sum$ is the summation sign, and boldface indicates vectors.

There are several forces acting together one way or the other in the atmosphere for the term on the right-hand side in Eq. (4.5). They are the pressure-gradient force, gravitational force, Coriolis force, and frictional force.

4.2.1   The Pressure-Gradient Force

The pressure-gradient force is caused by a gradient of pressure and is directed from high toward low pressure. Consider a small fluid block whose sides are dx, dy, and dz, as in Fig. 4.2. Let us calculate the pressure force on this fluid element in the x direction. The force due to the surrounding fluid on the left-hand vertical face, F_1, is $+p \, dy \, dz$, where the positive sign indicates that the force is directed to the right. The force on the right-hand vertical face, F_2, is

$$-[p + (\partial p/\partial x) \, dx] \, dy \, dz$$

where the additional term $(\partial p/\partial x) \, dx$ takes into account the variation of pressure through the small distance dx. The minus sign expresses the fact that this force is directed to the left. The net pressure force in the x direction is the sum of these two:

$$\text{Pressure force} = -(\partial p/\partial x) \, dx \, dy \, dz$$

or

$$\text{Pressure force per unit mass} = -\left(\frac{\partial P}{\partial x}\right) \frac{dx \, dy \, dz}{dM}$$

$$= -\left(\frac{\partial P}{\partial x}\right) \frac{dV}{dM}$$

$$= -\frac{1}{\rho}\left(\frac{\partial P}{\partial x}\right) \tag{4.6}$$

where dM, dV, and ρ are the mean mass, volume, and density of this infinitesimal cube, respectively. A similar equation can be derived for the y direction.

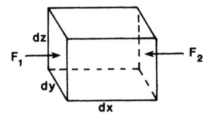

Fig. 4.2. Calculation of the pressure-gradient force.

4.2.2 The Coriolis Force

The Coriolis force arises owing to the earth's rotation. It may be easier to understand by following the description of Williams *et al.* (1968).

Imagine a target on the equator and a gun crew at the North Pole (Fig. 4.3). Assume that the gun crew can see the target and can aim their gun accordingly. The gun crew fires their gun, but since the target is moving at a different linear speed than the gun, they will miss the target. As a matter of fact, if the projectile had a speed of 1 km/sec, by the time it reached the equator it would miss the

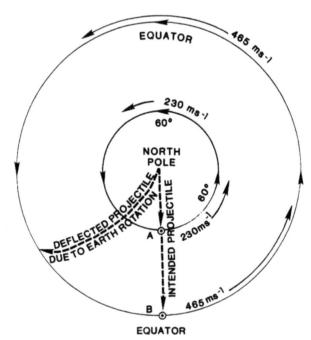

Fig. 4.3. The Coriolis effect.

target by something like 4600 km, a little more than one-eighth the equatorial circumference. Now it is obvious why the gun crew missed their target: the target was moving and they did not allow for this motion. But to the gun crew it would appear that there was a force acting on their projectile, tending to deflect it to the right. As far as they could see, the target was not moving with respect to them, since both the gun and the target were on the same surface rotating together. It may similarly be shown that a projectile fired from the equator toward the North Pole would again undergo a deflection to the right.

If one desires to try this same experiment on a smaller scale, he might attempt to play catch on a carnival merry-go-round while it is in motion. He would find that it would be extremely difficult to throw the ball so that it could be caught easily, for the same basic reason that the gun crew had difficulty. The linear speed of two different points on a rotating surface will be different simply because of rotation. This apparent deflection of a moving body, owing to the rotation of the earth, is called the Coriolis deflection, and the force that produces the deflection is called the Coriolis force.

The Coriolis force has been formally derived by, for example, Hess (1959). Mathematically, the horizontal components of this force are $2\Omega v \sin \phi$ due to northward motion, and $-2\Omega u \sin \phi$ due to eastward motion, where Ω is the angular velocity of the earth, ϕ is the latitude, and u and v represent east and north components of the velocity. Commonly,

$$f = 2\Omega \sin \phi \qquad (4.7)$$

is called the Coriolis parameter. Therefore, the Coriolis force in the x direction is fv, and in the y direction, $-fu$. Note that the Coriolis force acts as a "deflecting force," normal to the velocity, to the right of motion in the northern hemisphere (Fig. 4.3) and to the left in the southern hemisphere.

4.2.3 The Frictional Force

The frictional force, or simply the friction, is the mechanical resistive force offered by one medium or body to the relative motion of another medium or body in contact with the first (Huschke, 1959). If we assume that friction acts exactly opposite to the direction of motion, then a balance of the three forces becomes possible (Fig. 4.4). The pressure-gradient force will not be balanced by the Coriolis term alone, but by the vectorial sum of the frictional force and the Coriolis force. This requires the wind to blow at an angle α to the isobars (lines of equal pressure) toward low pressure.

Throughout most of the atmosphere, frictional forces are sufficiently small that, to a first approximation, they can be neglected (Wallace and Hobbs,

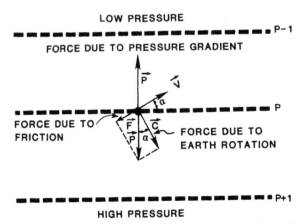

Fig. 4.4. Balance among pressure gradient, Coriolis, and friction forces in the northern hemisphere.

1977). A notable exception is the so-called planetary boundary layer (corresponding roughly to the lowest 1 km of the atmosphere), where the flow over a stationary underlying surface gives rise to a frictional drag force that is comparable in magnitude to the other terms in the horizontal equations of motion. For our purposes, it will be sufficient to represent this frictional drag in the highly simplified form (see, e.g., Wallace and Hobbs, 1977)

$$\mathbf{F} = -k\mathbf{V} \tag{4.8}$$

where k is a positive coefficient, the magnitude of which varies with wind speed, roughness of the underlying surface, static stability, and so on.

From Fig. 4.4 it is evident that

$$\mathbf{F} = \mathbf{P}\sin\alpha$$

and

$$\mathbf{P}\cos\alpha = \mathbf{C} = f\mathbf{V}$$

Combining these relationships, we obtain

$$\mathbf{F} = f\mathbf{V}\tan\alpha$$

It follows from Eq. (4.8) that

$$k = f\tan\alpha \tag{4.9}$$

In middle latitudes, $f \simeq 10^{-4}\ \mathrm{s}^{-1}$ if $\alpha \simeq 30°$ over the land near the surface, yielding a value of $k \simeq 0.6 \times 10^{-4}\ \mathrm{s}^{-1}$; therefore $k \simeq 60\%$ of parameter f. One thus cannot neglect the friction effect near the surface.

On the basis of Fig. 4.4 and Eq. (4.5), the horizontal equation of motion is therefore

$$dV/dt = P + C + F \tag{4.10}$$

and its x (eastward) and y (northward) components are

$$\frac{du}{dt} = -\frac{1}{\rho}\left(\frac{\partial P}{\partial x}\right) + fv - ku \tag{4.11}$$

$$\frac{dv}{dt} = -\frac{1}{\rho}\left(\frac{\partial P}{\partial y}\right) - fu - kv \tag{4.12}$$

The vertical component of the equation of motion can be derived (see, e.g., Hess, 1959) as

$$dw/dt = -(1/\rho)(\partial P/\partial z) - g - kw \tag{4.13}$$

where w is the vertical velocity and g is the gravitational acceleration. Note that if w is small and neglected, Eq. (4.13) reduces to the hydrostatic equation, as shown in Eq. (3.29).

4.3 Horizontal Motion under Balance of Forces

4.3.1 Geostrophic Flow

If one assumes that the motion is horizontal and there is no acceleration and no frictional effect, Eqs. (4.11), (4.12), and (4.13) become

$$fv = (1/\rho)\partial P/\partial x \tag{4.14}$$

$$fu = (-1/\rho)\partial P/\partial y \tag{4.15}$$

$$\partial P/\partial z = -\rho g \tag{4.16}$$

Note that Eq. (4.16) is the hydrostatic equation [see Eq. (3.29)].

Equations (4.14) and (4.15) express the balance between the pressure gradient and Coriolis forces per unit mass. The wind under this balance is called geostrophic flow (Fig. 4.5). Note that the geostrophic wind blows parallel to the isobars with low pressure on the left in the northern hemisphere and on the right in the southern hemisphere (because the Coriolis parameter f is negative). This behavior is known as Buys-Ballot's law. The total geostrophic wind velocity is $c = (u^2 + v^2)^{1/2}$.

Because parameter f becomes very small in the tropics, this geostrophic approximation is not good. Also, because the frictional effect is large near the

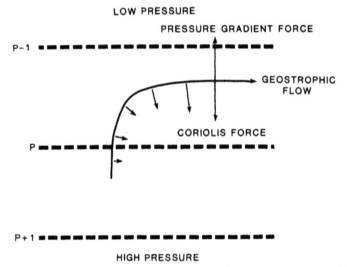

Fig. 4.5. Balance between pressure gradient and Coriolis forces to produce geostrophic flow in the northern hemisphere.

surface, this approximation cannot be applied. As depicted in Fig. 4.4, because of frictional effect the wind is not parallel to the isobar, but it blows toward the low pressure with an angle α. In general, however, the geostrophic wind is a good approximation.

4.3.2   Gradient Flow

Since isobars are necessarily straight lines, as assumed in previous sections, the flow along a curved path is subjected to the centrifugal force (see Fig. 4.6). From physics, this force is given as

$$C_e = \frac{c^2}{\gamma} \tag{4.17}$$

where C_e represents the centrifugal force per unit mass, c is the velocity, and γ is the radius of curvature.

If one assumes horizontal motion, no acceleration, and no frictional effect, one has for a cyclone (see Fig. 4.6).

$$P = C_e + C_o$$

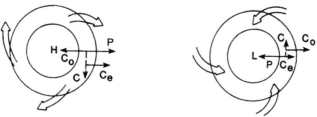

Fig. 4.6. Flow around curved paths in the northern hemisphere, where P stands for pressure gradient force, C_e for centrifugal force, C_o for Coriolis force, C for wind, H for high-pressure center, and L for low-pressure center. Note that large arrows around H and L represent the resultant wind direction caused by frictional effect.

Since P is $-(1/\rho)(\partial P/\partial \rho)$ [see, e.g., Eq. (4.6)] and C_o is fC [Eq. (4.7)], we have

$$-\frac{1}{\rho}\frac{\partial P}{\partial \gamma} = \frac{C^2}{\gamma} + fC$$

or

$$\frac{C^2}{\gamma} + fC + \frac{1}{\rho}\left(\frac{\partial P}{\partial \gamma}\right) = 0 \qquad (4.18)$$

Similarly, for an anticyclone,

$$P + C_e = C_o$$

or

$$-\frac{1}{\rho}\left(\frac{\partial P}{\partial \gamma}\right) + \frac{C^2}{\gamma} = fC$$

or

$$\frac{C^2}{\gamma} - fC - \frac{1}{\rho}\left(\frac{\partial P}{\partial \gamma}\right) = 0 \qquad (4.19)$$

Equations (4.18) and (4.19) are simple quadratics. The solutions are, for cyclones,

$$C = -\frac{f\gamma}{2} + \left(\frac{f^2\gamma^2}{4} - \frac{\gamma}{\rho}\frac{\partial P}{\partial \gamma}\right)^{1/2} \qquad (4.20)$$

and for anticyclones,

$$C = \frac{f\gamma}{2} - \left(\frac{f^2\gamma^2}{4} + \frac{\gamma}{\rho}\frac{\partial P}{\partial \gamma}\right)^{1/2} \qquad (4.21)$$

Note that conditions when $\partial P/\partial y = 0$, $C = 0$, are applied in the derivation of these equations.

It can be shown (see Hess, 1959, p. 186) that in a low the geostrophic value is an overestimate of the gradient wind, whereas in normal flow around a high the geostrophic value is an underestimate of the gradient wind speed. In practice, however, the geostrophic wind, in most cases, is a better approximation to the observed value than the gradient wind.

4.3.3 Cyclostrophic Flow

As shown in Fig. 1.1, there are small-scale systems such as tornadoes, water spouts, dust devils, hurricanes, and typhoons. Their radius of rotation is much smaller than, say, synoptic scales such as high- and low-pressure systems. Under these conditions, the centrifugal force, which is proportional to C^2, is much larger than the Coriolis force, which is proportional to C instead of C^2. Thus, pressure gradient force may be balanced with the centrifugal force so that

$$\frac{C^2}{\gamma} = \frac{1}{\rho}\left(\frac{\partial P}{\partial \gamma}\right) \tag{4.22}$$

This type of motion is called cyclostrophic flow, which may be used to estimate the wind speed around a small but highly concentric motion if γ and $\partial P/\partial y$ are measured.

4.3.4 The Thermal Wind Equation

The thermal wind equation is an equation for the vertical variation of the geostrophic wind under hydrostatic equilibrium conditions. It can be derived (see Hess, 1959, p. 191) from geostrophic wind equations,

$$fv = \frac{1}{\rho}\left(\frac{\partial P}{\partial x}\right) \quad \text{and} \quad fu = -\frac{1}{\rho}\left(\frac{\partial P}{\partial y}\right)$$

and from the hydrostatic equation and the equation of state,

$$g = -\frac{1}{\rho}\left(\frac{\partial P}{\partial z}\right) \quad \text{and} \quad \rho = \frac{P}{RT}$$

Elimination of ρ in the above equations gives

$$\frac{fv}{T} = R\frac{\partial \ln P}{\partial x} \tag{4.23}$$

$$\frac{fu}{T} = -R\frac{\partial \ln P}{\partial y} \tag{4.24}$$

$$\frac{g}{T} = -R\frac{\partial \ln P}{\partial z} \tag{4.25}$$

Differentiation of Eq. (4.23) with respect to z gives

$$f\frac{\partial}{\partial z}\left(\frac{v}{T}\right) = R\frac{\partial}{\partial z}\frac{\partial \ln P}{\partial x}$$

or

$$\frac{f}{T^2}\left(T\frac{\partial v}{\partial z} - v\frac{\partial T}{\partial z}\right) = R\frac{\partial}{\partial z}\frac{\partial \ln P}{\partial x} \tag{4.26}$$

Differentiation of Eq. (4.25) with respect to x gives

$$-\frac{g}{T^2}\frac{\partial T}{\partial x} = -R\frac{\partial}{\partial z}\frac{\partial \ln P}{\partial x} \tag{4.27}$$

Addition of Eqs. (4.26) and (4.27) gives

$$\frac{\partial v}{\partial z} = \frac{g}{fT}\left(\frac{\partial T}{\partial x}\right) + \frac{v}{T}\left(\frac{\partial T}{\partial z}\right) \tag{4.28}$$

Similarly, differentiating Eqs. (4.24) and (4.25) with respect to z and y, respectively, and eliminating the pressure term gives

$$\frac{\partial u}{\partial z} = -\frac{g}{fT}\left(\frac{\partial T}{\partial y}\right) + \frac{u}{T}\left(\frac{\partial T}{\partial z}\right) \tag{4.29}$$

Since the second terms on the right in Eqs. (4.28) and (4.29) are small, as reasoned by Hess (1959, p. 191), one may write the thermal wind equations as

$$\frac{\partial v}{\partial z} \simeq \frac{g}{fT}\left(\frac{\partial T}{\partial x}\right) \qquad \frac{\partial u}{\partial z} \simeq -\frac{g}{fT}\left(\frac{\partial T}{\partial y}\right) \tag{4.30}$$

These equations require that, in the northern hemisphere, v increase with height if the temperature increases to the east and u increase with height if $\partial T/\partial y$ is negative or the temperature increases to the south. Thus, the thermal wind is directed along the isotherms, with cold air to the left in the northern hemisphere and to the right in the southern hemisphere.

It can be shown (Hess, 1959, p. 193) that the wind turns anticyclonically with height (veers) when there is a wind component from warm toward cold air, and the wind turns cyclonically with height (backs) whenever there is a wind component from cold toward warm air. In other words, veering of the geostrophic wind with height occurs with warm-air advection, and backing of the geostrophic wind with height occurs with cold-air advection.

4.4   The Equation of Motion in Turbulent Flow

4.4.1   The Eddy Viscosity

The effect of eddy viscosity is illustrated in Fig. 4.7. As the wind blows over the sea surface it exerts a certain force F upon the surface to keep it moving. This force is proportional to the area of the surface and to the speed v. It is inversely proportional to the distance z separating the surface and the sea bed. Therefore

$$F = \mu \frac{AV}{z}$$

or
$$\tau = \frac{F}{A} = \mu \frac{\partial V}{\partial z} \tag{4.31}$$

where μ is a constant of proportionality, called the coefficient of eddy viscosity, and τ is the wind stress or applied force per unit area. Note that in the steady state the variation of speed is linear, so that F is also proportional to the derivative of V with respect to z.

For a cube of fluid, there are nine possible stresses. For example, in the y direction we have τ_{xy}, τ_{yy}, and τ_{zy}, where τ_{xy} means a force per unit area on a face normal to the x axis due to motion in the y direction. Note that the stress term must be added to the appropriate equation of motion if the frictional effect is taken into account. For example, the y equation of motion becomes (see Hess, 1959, p. 268).

$$\frac{dv}{dt} = -fu - \frac{1}{\rho}\left(\frac{\partial P}{\partial v}\right) + \frac{1}{\rho}\left(\frac{\partial \tau_{xy}}{\partial x} + \frac{\partial \tau_{yy}}{\partial y} + \frac{\partial \tau_{zy}}{\partial z}\right) \tag{4.32}$$

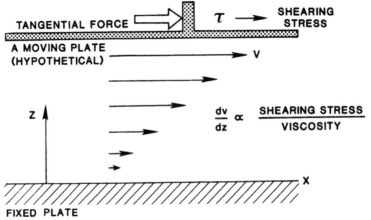

Fig. 4.7. A schematic of fluid shear.

4.4.2   Eddy Velocities

The mean value and eddy velocities are illustrated in Fig. 4.8, which represents wind fluctuation in the x direction. Let u, v, w be the components of velocity measured at a point (x, y, z). In turbulent flow all three components are functions of time as well as of position. The mean velocity, with components $\bar{u}, \bar{v}, \bar{w}$, is defined at a fixed point and at a time t_0 by the relation (see, e.g., Sutton, 1953)

$$\bar{u} = \frac{1}{T} \int_{t_0 - 1/2T}^{t_0 + 1/2T} u\, dt$$

$$\bar{v} = \frac{1}{T} \int_{t_0 - 1/2T}^{t_0 + 1/2T} v\, dt$$

$$\bar{w} = \frac{1}{T} \int_{t_0 - 1/2T}^{t_0 + 1/2T} w\, dt \tag{4.33}$$

where T is called the period of sampling.

The eddy velocity such as $u', v',$ or w' is defined as the difference between the total velocity at any instant, u, and the mean velocity, such that

$$u' = u - \bar{u} \qquad v' = v - \bar{v} \qquad w' = w - \bar{w} \tag{4.34}$$

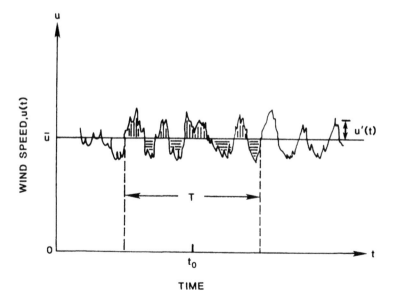

Fig. 4.8. Definition of mean and fluctuation or eddy velocities.

In steady mean flow

$$\bar{u} = \bar{\bar{u}}, \text{ etc.}$$

and

$$\bar{u}' = \frac{1}{T} \int_{t_0 - 1/2T}^{t_0 + 1/2T} u' \, dt$$

$$= \frac{1}{T} \int_{t_0 - 1/2T}^{t_0 + 1/2T} (u - \bar{u}) \, dt$$

$$= \bar{u} - \bar{u} = 0$$

Thus

$$\bar{u}' = \bar{v}' = \bar{w}' = 0$$

Note that terms such as $\overline{u'^2}$, $\overline{u'w'}$, etc., are not necessarily zero. For example,

$$\overline{vv} = \overline{(\bar{v} + v')(\bar{v} + v')} = \bar{v}\bar{v} + \overline{v'v'} \tag{4.35}$$

where terms such as $\overline{v'v'}$ do not vanish if there is a nonzero correlation between the two departures whose product is averaged.

4.4.3   The Equations of Mean Motion in Turbulent Flow

The equations of motion for instantaneous motion, for example, in the y direction, may be written [cf. Eq. (4.12) without friction term]

$$\frac{dv}{dt} = -fu - \frac{1}{\rho}\left(\frac{\partial P}{\partial y}\right)$$

or

$$\rho\frac{\partial v}{\partial t} + \rho u\frac{\partial v}{\partial x} + \rho v\frac{\partial v}{\partial y} + \rho w\frac{\partial v}{\partial z} = -\rho fu - \frac{\partial P}{\partial y} \tag{4.36}$$

From Eq. (4.4), the equation of continuity may be written

$$\rho v\frac{\partial u}{\partial x} + \rho v\frac{\partial v}{\partial y} + \rho v\frac{\partial w}{\partial z} = 0 \tag{4.37}$$

Equation (4.36) may be added to Eq. (4.37) so that

$$\frac{\partial(\rho v)}{\partial t} + \frac{\partial(\rho vu)}{\partial x} + \frac{\partial(\rho vv)}{\partial y} + \frac{\partial(\rho vw)}{\partial z} = -\rho fu - \frac{\partial P}{\partial y}$$

For the mean of each term, one has

$$\frac{\partial(\rho\bar{v})}{\partial t} + \frac{\partial(\rho\overline{vu})}{\partial x} + \frac{\partial(\rho\overline{vv})}{\partial y} + \frac{\partial(\rho\overline{vw})}{\partial z} = -\rho f\bar{u} - \frac{\partial\bar{P}}{\partial y}$$

Applying Eq. (4.35), one has

$$\frac{\partial(\rho\bar{v})}{\partial t} + \frac{\partial(\rho\overline{vu})}{\partial x} + \frac{\partial(\rho\overline{v'u'})}{\partial x} + \frac{\partial(\rho\overline{vv})}{\partial y} + \frac{\partial(\rho\overline{v'v'})}{\partial y}$$

$$+ \frac{\partial(\rho\overline{vw})}{\partial z} + \frac{\partial(\rho\overline{v'w'})}{\partial z} = -\rho f\bar{u} - \frac{\partial\bar{P}}{\partial y}$$

or

$$\rho\frac{\partial\bar{v}}{\partial t} + \bar{v}\frac{\partial\rho}{\partial t} + \rho\bar{v}\frac{\partial\bar{u}}{\partial x} + \rho\bar{u}\frac{\partial\bar{v}}{\partial x}$$

$$+ \rho\bar{v}\frac{\partial\bar{v}}{\partial y} + \rho\bar{v}\frac{\partial\bar{v}}{\partial y} + \rho\bar{v}\frac{\partial\bar{w}}{\partial z} + \rho\bar{w}\frac{\partial\bar{v}}{\partial z}$$

$$= -\rho f\bar{u} - \frac{\partial\bar{P}}{\partial y} - \frac{\partial(\rho\overline{v'u'})}{\partial x} - \frac{\partial(\rho\overline{v'v'})}{\partial y} - \frac{\partial(\rho\overline{v'w'})}{\partial z} \qquad (4.38)$$

Since the averaged equation of continuity where multiplied by $\bar{v}$ is

$$\bar{v}\frac{\partial\bar{u}}{\partial x} + \bar{v}\frac{\partial\bar{v}}{\partial y} + \bar{v}\frac{\partial\bar{w}}{\partial z} = 0$$

Eq. (4.38) becomes

$$\frac{\partial\bar{v}}{\partial t} + \bar{u}\frac{\partial\bar{v}}{\partial x} + \bar{v}\frac{\partial\bar{v}}{\partial y} + \bar{w}\frac{\partial\bar{v}}{\partial z}$$

$$= -f\bar{u} - \frac{1}{\rho}\left(\frac{\partial\bar{P}}{\partial y}\right) - \left[\frac{1}{\rho}\frac{\partial(\rho\overline{v'u'})}{\partial x} + \frac{\partial(\rho\overline{v'v'})}{\partial y} + \frac{\partial(\rho\overline{v'w'})}{\partial z}\right] \qquad (4.39)$$

This is the equation of mean motion in the y direction for a turbulent incompressible fluid.

Comparison between Eq. (4.39) and respective mean terms in Eq. (4.32) indicates that

$$\tau_{xy} = -\rho\overline{v'u'}$$

$$\tau_{yy} = -\rho\overline{v'v'}$$

$$\tau_{zy} = -\rho\overline{v'w'} \qquad (4.40)$$

According to Eq. (4.23), one also gets

$$-\overline{\rho v'w'} = \mu\frac{\partial\bar{v}}{\partial z} \tag{4.41}$$

for the y direction, or

$$\tau_{zx} = -\overline{\rho u'w'} = \mu\frac{\partial\bar{u}}{\partial z} \tag{4.42}$$

for the x direction, which is usually taken as along the prevailing wind direction.

From Fig. 4.7 and Eq. (4.42), one can measure the wind stress on the sea surface by direct measurements of the fluctuation of u' and w', the so-called eddy correlation method. More about this will be given in Chapter 6.

4.4.4   Numerical Indices

In meteorology and oceanography, there are several nondimensional ratios that are important in modeling and dynamic similitude. They can be derived by transforming the basic equations of motion to nondimensional form (Hess, 1959). The Reynolds number, Re, is the ratio of inertial force to the viscous force in fluid motion, that is,

$$\text{Re} = \frac{LU}{v} \tag{4.43}$$

where L is a characteristic length, v is the kinematic viscosity, and U is a characteristic velocity.

The Froude number, Fr, is the nondimensional ratio of the inertial force to the force of gravity for a given fluid flow, that is,

$$\text{Fr} = \frac{U^2}{Lg} \tag{4.44}$$

where g is the acceleration of gravity.

The Rossby number, Ro, is the nondimensional ratio of the inertial force to the Coriolis force for a given flow of a rotating fluid, that is,

$$\text{Ro} = \frac{U}{fL} \tag{4.45}$$

where f is the Coriolis parameter.

On the basis of turbulent diffusion of kinetic energy, the Richardson number can also be derived (see, e.g., Hess, 1959, p. 290). The Richardson number, Ri, is the nondimensional ratio of the thermal stability to the wind shear, that is,

$$\text{Ri} = \frac{(g/\bar{\theta})(\partial\bar{\theta}/\partial z)}{(\partial\bar{u}/\partial z)^2} \tag{4.46}$$

Since the denominator is always positive, the sign of Ri is dependent on the term $\partial\bar{\theta}/\partial z$ in the numerator. In other words, under stable conditions, where θ increases with height so that $\partial\bar{\theta}/\partial z$ is positive, Ri is positive. On the other hand, under unstable conditions where θ decreases with height such that $\partial\theta/\partial z$ is negative, then Ri is negative. Under adiabatic conditions, however, $\partial\theta/\partial z = 0$ and Ri = 0.

4.5 Some Kinematics of Fluid Flow

Although the kinematics of the atmosphere deals with the description of the motion of air without reference to the forces producing the motion, some of the kinematics are included here as a part of atmospheric dynamics.

A *streamline* is a line that, at any given instant, is tangent to all the velocity vectors of the points through which the line passes; thus the flow is along the streamlines at any given moment (Hess, 1959). Since in the tropics the Coriolis parameter is small, the geostrophic approximation does not work well. Streamline analysis is an effective way to study the air flow characteristics (see, e.g., Atkinson, 1971).

A *trajectory* is a line along which a given fluid particle has moved—for example, the path traced by a radioactive cloud (Hess, 1959).

Vorticity is a vector measure of local rotation in a fluid flow (Huschke, 1959). The term vorticity, that is,

$$\frac{\partial v}{\partial x} - \frac{\partial u}{\partial y} \quad \text{or} \quad \frac{\partial v}{\partial x} + \left(-\frac{\partial u}{\partial y}\right)$$

can be derived by Taylor's series expansion in a rotational coordinate system (Hess, 1959).

As shown in Fig. 4.9, both terms, $(\partial v/\partial x)$ and $(-\partial u/\partial y)$, make a positive contribution in a counterclockwise motion. Thus cyclonic motion is associated with positive vorticity, and anticyclonic motion with negative vorticity.

Circulation is a precise measure of the average flow of fluid along a given closed curve (Huschke, 1959). It is the line integral

$$C = \oint (u\, dx + v\, dy + w\, dz) \tag{4.47}$$

about the closed curve.

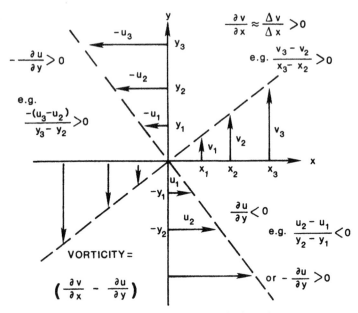

Fig. 4.9. Positive vorticity and cyclonic motion.

The relationship between circulation and vorticity is (see, e.g., Hess, 1959)

$$\oint (u\, dx + v\, dy) = \iint \left(\frac{\partial v}{\partial x} - \frac{\partial u}{\partial y} \right) dx\, dy \qquad (4.48)$$

where the left side is the circulation around the given curve and the right side is the integral of vorticity over the area enclosed by the curve. Thus, circulation is an areal measure of the rotational tendency of a fluid and vorticity is a point measure of that same tendency. Equation (4.48) shows that if there is no circulation there will be no vorticity. Thus the fluid is said to be irrotational.

Chapter 5 | Synoptic Meteorology

Synoptic meteorology is the use of meteorological data obtained simultaneously over a wide area for the purpose of presenting a comprehensive and nearly instantaneous picture of the state of the atmosphere.

In order to understand meteorological phenomena in the coastal zone, larger-scale and synoptic conditions must be appreciated. Because many textbooks on synoptic meteorology are available, the subject will be discussed in general terms only and emphasis will be placed on effects in coastal zones.

5.1 The General Circulation

General circulation refers to atmospheric motions over the earth. Figure 5.1 is a very much generalized wind, pressure, and vertical circulation pattern for the world. This pattern exists mainly because of the heat flow from the tropics to polar regions and the Coriolis effect on the deflection of winds. A schematic representation of the major characteristic atmospheric structures in the northern hemisphere is presented in Fig. 5.2

Because tropical regions receive more solar radiation (owing to direct sun rays) than other latitudes, particularly polar regions, the excess heat produces convective currents in the equatorial convergence zone. Following the equation of continuity, these low-level convergences are compensated by upper-level divergence. The excess heat is then transported to higher latitudes. However, the ascending air is colder than the surrounding air at high altitude, and it begins to sink. The descending motion produces subsidence and adiabatic warming. The subsiding air spreads out, some moving back toward the equator and some moving to higher latitudes. The center of the subsiding

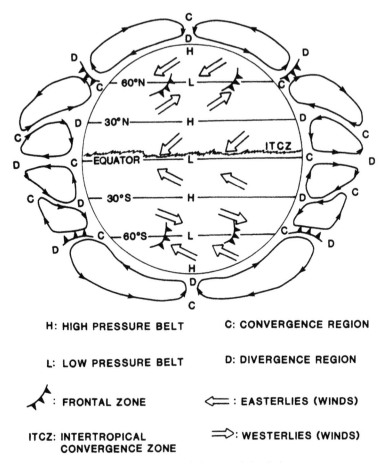

H: HIGH PRESSURE BELT   C: CONVERGENCE REGION

L: LOW PRESSURE BELT   D: DIVERGENCE REGION

⤲ : FRONTAL ZONE   ⇐ : EASTERLIES (WINDS)

ITCZ: INTERTROPICAL   ⇒ : WESTERLIES (WINDS)
CONVERGENCE ZONE

Fig. 5.1. Schematic of the general circulation.

air, along approximately 30°N and S, is called high pressure. Because of the Coriolis effect, northeast trade winds (called simply trades) prevail roughly between the equator and 30°N, and southeast trades between the equator and 30°S. Between 30 and 60°N or S, westerlies (i.e., wind blowing from the west) prevail. The descending motion that moves to high latitudes eventually meets the descending air from the polar region, producing an ascending current. The collision of the vertical motions is called the polar front zone. Subsidence over polar regions is produced principally because a permanent high-pressure system exists over these regions. Therefore, in polar regions easterlies prevail.

Circulations along the equator are shown in Fig. 5.3. Note that the dominating structure appears as rising motion over the Indonesian region and the warm sea surface of the west Pacific Ocean, with descent to the east and

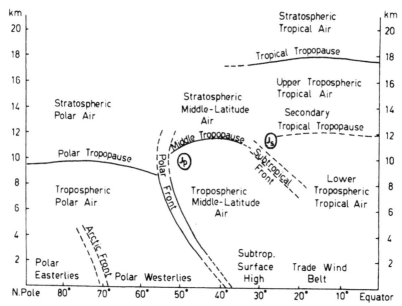

Fig. 5.2. Principal air masses, tropopauses and fronts, and jet streams in relation to features of low-level wind systems. Depending on location and time, fronts may be either well developed or weak. [After Palmen and Newton (1969).]

west. The Pacific cell associated with Indonesian convection, often called the "Walker circulation," spans the breadth of the Pacific Ocean and provides a physical manifestation of Walker's southern oscillation index (Webster, 1983).

The constancy of the surface wind direction versus latitude, expressed in percentages, is shown in Fig. 5.4. Note that the trade winds attain a constancy of 80% (Kotsch, 1983).

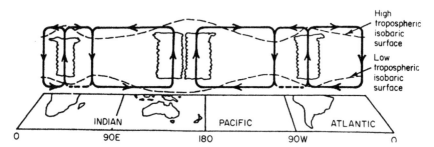

Fig. 5.3. Schematic view of the equatorial symmetric planetary scale features. Note the dominance of the Pacific Ocean–Indonesian cell, which is referred to as the Walker Circulation. [After Webster (1983). A modified version of this schematic by Webster and Chang will be published in the *Journal of Atmospheric Sciences* in 1988.]

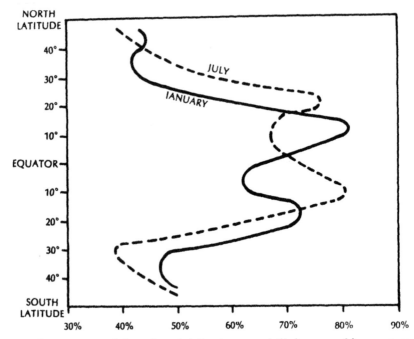

Fig. 5.4. Constancy of the surface-wind direction versus latitude expressed in percentages. Note that the trade winds attain a constancy of 80%. [After Kotsch (1983).]

For study of day-to-day weather maps, the above discussion is far too simple because it does not consider other effects such as ocean–continent contrast and topographic influence on the airflow.

5.2   Air Masses and Fronts

An air mass is often defined as a widespread body of air that is approximately homogeneous in its horizontal extent, particularly with reference to temperature and moisture distribution; in addition, vertical temperature and moisture variations are approximately the same over its horizontal extent (Huschke, 1959).

Major types of air masses and their characteristics are summarized in Table 5.1. It can be inferred that for a given region such as the Gulf Coast of the Unites States, one would experience more frequently cP air during the winter season and mT air during summer months.

An atmospheric front is the interface or transition zone between two air masses of different density. Since the temperature distribution is the most

Table 5.1

Major Air Mass Types

Source region	Abbreviation	Characteristics
Continental polar Origin: Land areas in high latitudes	cP	Air is colder and drier than in low latitudes
Continental tropical Origin: Land areas in low latitudes	cT	Warmer and drier than in higher latitudes
Maritime polar Origin: Oceanic areas in high latitudes	mP	Colder and moister than in low latitudes
Maritime tropical Origin: Oceanic areas in low latitudes	mT	Warmer and moister than in high latitudes

important regulator of atmospheric density, a front almost invariably separates air masses of different temperatures.

Depending on their direction of movement, fronts can be classified as cold, warm, occluded, and stationary. A cold front (Fig. 5.5a) exists when a cold air mass displaces a warm air mass all along the frontal zone. The average slope of the cold front ranges from 1/50 (height/horizontal distance) to 1/150. The average ground speed of the cold front is $10-15$ m s^{-1}. Heavy rains, gusty winds, and thunderstorms are usually associated with the passage of a cold front.

When a warm air mass displaces a cold air mass, a warm front prevails (Fig. 5.5b). The average slope of the warm front ranges from 1/100 to 1/300. The average ground speed is around 5 m s^{-1}. The rainy area is much broader than that produced by a cold front. Rain-producing clouds are generally of the stratiform type rather than convective clouds such as cumulonimbus, as in the cold front case.

An occluded front is generated when a cold front overtakes a warm front and lifts the warm air mass completely off the ground. According to Huschke (1959), there are three basic types of occluded front, determined by the relative coldness of the air behind the original cold front compared to the air ahead of the warm (or stationary) front. (a) A cold occlusion (Fig. 5.5c) results when the coldest air is behind the cold front. The cold front undercuts the warm front and, at the earth's surface, coldest air replaces less cold air. (b) When the coldest air lies ahead of the warm front, a warm occlusion (Fig. 5.5d) is formed, in which case the original cold front is forced aloft at the warm front surface. At the earth's surface, coldest air is replaced by less cold air. (c) A third and

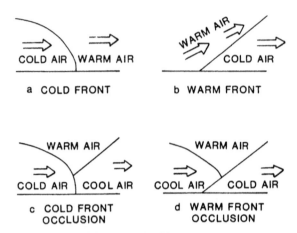

Fig. 5.5. Schematic of frontal systems.

frequent type, a neutral occlusion, results when there is no appreciable temperature difference between the cold air masses of the cold and warm fronts. In this case frontal characteristics at the earth's surface consist mainly of a pressure trough, a wind shift line, and a band of cloudiness and precipitation.

When there is no appreciable motion of the front, the transition zone between the two air masses commonly is called a quasi-stationary front, or stationary front. Frontal overrunning is normally associated with a stationary front. Overrunning is a condition existing when an air mass is in motion above another air mass of greater density at the surface. This term usually is applied in the case of warm air ascending the surface of a warm front or stationary front (Huschke, 1959).

Along the U.S. Gulf Coast or over the northern Gulf of Mexico, frontal overrunning occurs frequently when a polar front is nearly stationary (see, e.g., Muller, 1977). A polar front is the semipermanent, semicontinuous front separating air masses of tropical and polar origin. Under overrunning conditions, heavy cloud cover and precipitation are usually associated with this kind of weather system and may cause operational problems such as disruption of onshore–offshore helicopter flights for offshore oil field workers and supplies because of stronger wind shear and lower visibility than normal.

Since frontal overrunning is important from an operational standpoint, its frequency of occurrence may be studied by estimating the slope of the stationary front (Hsu, 1981a). A schematic is shown in Fig. 5.6. The pressures at points 2 and 4 are

$$P_2 = P_3 + \rho_{\text{cold}} g \, \Delta Z \tag{5.1}$$

$$P_4 = P_1 - \rho_{\text{warm}} g \, \Delta Z \tag{5.2}$$

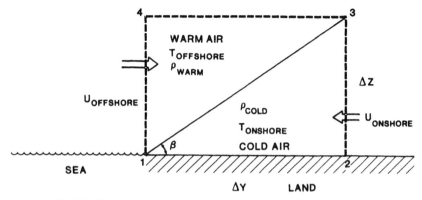

Fig. 5.6. Schematic of the slope of a stationary front in the coastal zone.

Equations (5.1) and (5.2) may be rewritten as

$$P_2 - P_1 = P_3 - P_1 + \rho_{cold} g \, \Delta Z \tag{5.3}$$

$$P_3 - P_4 = P_3 - P_1 + \rho_{warm} g \, \Delta Z \tag{5.4}$$

Subtracting Eq. (5.4) from Eq. (5.3), one gets

$$(P_2 - P_1) - (P_3 - P_4) = \rho_{cold} g \, \Delta Z - \rho_{warm} g \, \Delta Z$$

Dividing the above equation by Δy and rearranging, we have

$$\frac{\Delta Z}{\Delta Y} = \tan \beta = \frac{[(P_2 - P_1)/\Delta Y] - [(P_3 - P_4)/\Delta Y]}{g(\rho_{cold} - \rho_{warm})} \tag{5.5}$$

where $\Delta Z/\Delta Y$ represents the slope of the front.
   Under geostrophic flow conditions [cf. Eq. (4.15)],

$$\frac{P_2 - P_1}{\Delta Y} = -\rho_{cold} f u_{onshore} \tag{5.6a}$$

$$\frac{P_3 - P_4}{\Delta Y} = -\rho_{warm} f u_{offshore} \tag{5.6b}$$

Substituting Eq. (5.6) into Eq. (5.5), one gets

$$\tan \beta = \frac{-\rho_{cold} f u_{onshore} + \rho_{warm} f u_{offshore}}{g(\rho_{cold} - \rho_{warm})} \tag{5.7}$$

From the equation of state (Fig. 3.3),

$$P_{warm} = \rho_{warm} R T_{warm}$$

$$P_{cold} = \rho_{cold} R T_{cold} \tag{5.8}$$

Substituting Eq. (5.8) into Eq. (5.7), and assuming that at any point on the frontal surface $P_{cold} = P_{warm}$, we have

$$\tan\beta = \left(\frac{f}{g}\right)\frac{T_{cold}u_{offshore} - T_{warm}u_{onshore}}{T_{warm} - T_{cold}}$$

Taking the mean of the two temperatures in the numerator, T_m (see, e.g., Byers, 1974, p. 244), we have

$$\tan\beta = \left(\frac{fT_m}{g}\right)\frac{u_{offshore} - u_{onshore}}{T_{offshore} - T_{onshore}} \tag{5.9}$$

where $T_{onshore} = T_{cold}$ and $T_{offshore} = T_{warm}$.

Figure 5.7 shows that over the central U.S. Gulf Coast the percentage of occurrence of monthly frontal overrunning is controlled predominantly by temperature differences across the coastal zone and to a lesser extent by wind speed differences. Further correlation between the overrunning percentage and temperature differences is shown in Fig. 5.8. The correlation coefficient (r) is 0.91. Considering only 12-pair (month) data points (Fig. 5.8), the correlation coefficient is statistically highly significant.

Meteorologically speaking, it is understood, and proved as formulated above, that the larger the temperature difference as shown in the denomina-

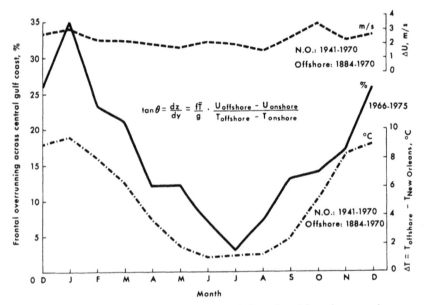

Fig. 5.7. Monthly variations in air temperature, wind speed, and frontal overrunning across the central Gulf Coast (Hsu, 1981a).

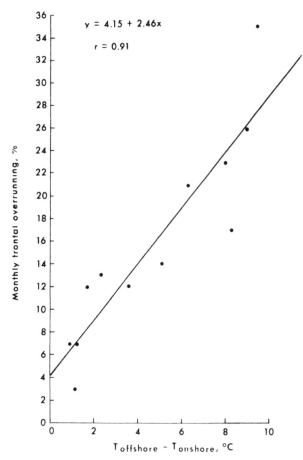

Fig. 5.8. Relationship between monthly frontal overrunning and offshore-onshore temperature differences across the central Gulf Coast (Hsu, 1981a).

tor in Eq. (5.9), the smaller the frontal slope and therefore the more chance there is for overrunning to occur. This is shown by the solid curve in Fig. 5.7, which essentially relates to the slope or the left-hand side of Eq. (5.9). Since there is very little variation in the wind speed difference (Fig. 5.7) from month to month as compared to temperature differences, the numerator in Eq. (5.9) contributes rather insignificantly to the frontal slope. If this reasoning is accepted, it is possible to use the monthly temperature difference between onshore stations such as New Orleans and offshore measurements made by meteorological satellites or buoy or rig observations to forecast the probability of local overrunning. This approach may be useful for environmental planning.

5.3 Extratropical Cyclones and Cyclogenesis

The term "cyclonic" means having the same sense of rotation about the local vertical as the earth's rotation, that is, as viewed from above, counterclockwise in the northern hemisphere, clockwise in the southern hemisphere, and undefined at the equator. The opposite is anticyclonic, which means having a sense of rotation about the local vertical opposite to that of the earth's rotation; that is, clockwise in the northern hemisphere, counterclockwise in the southern hemisphere, and undefined at the equator. A cyclone is a closed-circulation system having cyclonic motion. Because cyclonic circulation and relatively low atmospheric pressure usually coexist, in common practice the terms "cyclone" and "low" (elliptical for area of low pressure) are used interchangeably. Also, because cyclones nearly always are accompanied by inclement (often destructive) weather, they are frequently referred to simply as storms. As a closed-circulation system having anticyclonic motion, that is, anticyclonic circulation, and relatively high atmospheric pressure usually coexist, the terms "anticyclone" and "high" (elliptical for area of high pressure) are used interchangeably in common practice (Huschke, 1959).

Extratropical cyclones are those cyclones associated with migratory fronts occurring in middle and high latitudes. Schematics of the life cycles of extratropical cyclones in the northern and southern hemispheres are illustrated in Figs. 5.9 and 5.10, respectively. In the beginning, a small wavelike

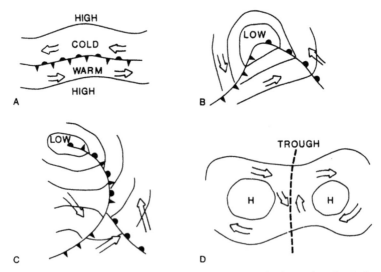

Fig. 5.9. Schematic of the life cycle of an extratropical cyclone in the northern hemisphere (for legend, see frontal symbols as listed in Appendix E).

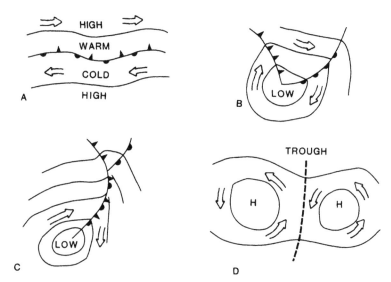

Fig. 5.10. Schematic of the life cycle of an extratropical cyclone in the northern hemishpere (for legend, see frontal symbols as listed in Appendix E).

indentation along a stationary front is formed (see Fig. 5.9A). If the condition intensifies, a cyclonic circulation will appear around the low that occurs at the apex of the wave. In the process, both cold and warm fronts will be formed (Fig. 5.9B). If the cold front moves faster than and "catches up" with the warm front, an occlusion front will be formed (Fig. 5.9C). As the process continues, the warm air is displaced aloft. Eventually the center of the cyclone will fill with cold air, which is completely underneath warm air. A series of weak low-pressure systems, or a trough, will then prevail (Fig. 5.9D). Note that usually a series of cyclones or so-called "cyclone families" along the polar front will occur instead of just one cyclone by itself. This can be seen in daily middle- to high-latitude weather maps.

Cyclogenesis is a very important phenomenon in certain coastal regions. Cyclogenesis is defined as any development or strengthening of cyclonic circulation in the atmosphere; the opposite is cyclolysis. The term is applied to the development of cyclonic circulation where previously it did not exist (commonly, the initial appearance of a low or trough, as well as the intensification of existing cyclonic flow) (Huschke, 1959).

Cyclogenesis is not unusual in certain coastal regions, for example, along the mid-Atlantic coast of the United States and in the northwestern Gulf of Mexico. According to Miller (1946), cyclones that form off the mid-Atlantic coast can be divided into two types. One of these develops well offshore along or east of the Gulf Stream and moves in a northeasterly direction. The other originates in the Gulf coastal region or near the southeast coast and rapidly intensifies as it moves up the East Coast. This type generally develops as a secondary disturbance to a primary cyclone moving through the Great Lakes region. These secondary disturbances can give rise to devastating weather. One example is the President's Day snowstorm of February 18–19, 1979, which deposited a record-breaking snowfall on the Middle Atlantic states (Bosart, 1981; Uccellini *et al.*, 1984). The following discussion is based on Bosart (1981) and Uccellini *et al.* (1984).

A synoptic overview of the President's Day cyclone of February 18–19, 1979, is given in Fig. 5.11 (Uccellini *et al.*, 1984). At 12Z/17 [1200 Greenwich Meridian Time (GMT) February 17], a massive high-pressure system centered over Lake Superior (Fig. 5.11A) was marked by record low surface temperatures from southern Canada to the eastern United States. At 00Z/18 (Fig. 5.11B) the large surface anticyclone remained over the Great Lakes. A high-pressure ridge moved southward along the East Coast, reflecting the "damming" of cold air east of the Appalachian Mountains. Snow and sleet had developed over the southern United States with the first indication of an inverted trough extending northward from the Gulf of Mexico. By 12Z/18 (Fig. 5.11C) moderate to heavy snow had developed in the southern Appalachians as the inverted surface trough extended northward from the Gulf of Mexico and became more pronounced, while a separate trough began to develop off the southeast coast. By 00Z/19 (Fig. 5.11D) the low-pressure system that had previously been located in the central Gulf Coast had moved to the offshore regions of the Carolinas. A region of heavy snowfall was located along the East Coast. By 12Z/19 (Fig. 5.11E), rapid cyclogenesis was in progress along the Virginia coast, with heavy snow occurring from northern Virginia to extreme southeastern New York. Bosart's (1981) analysis shows that the coastal front was still present in eastern Maryland, providing the low-level convergence that enhanced snowfall rates from Washington, D.C., northward toward New York City. By 00Z/20 (Fig. 5.11F), the cyclone had developed into a major vortex and moved eastward, and its effect on the coastal regions thus diminished.

A detailed mesoscale sectional view of this developing coastal cyclone is provided in Fig. 5.12 by Bosart (1981). Broad onshore flow is seen at 1200 GMT on February 18, with a well-defined wind shift extending from just north of Daytona Beach, Florida, to the area east of the Georgia coast. The ship data are rather persuasive in this respect, with the temperature contrast averaging $10°C (100 \text{ km}^{-1})$ across the incipient front. In the following 6 hr the coastal front continued to increase in intensity, with a strong convergence zone extending well northward to off the North Carolina coast. By 0000 GMT on February 19 a closed cyclonic circulation is evident, with continued strong thermal contrast in the coastal zone. The developing cyclone center is

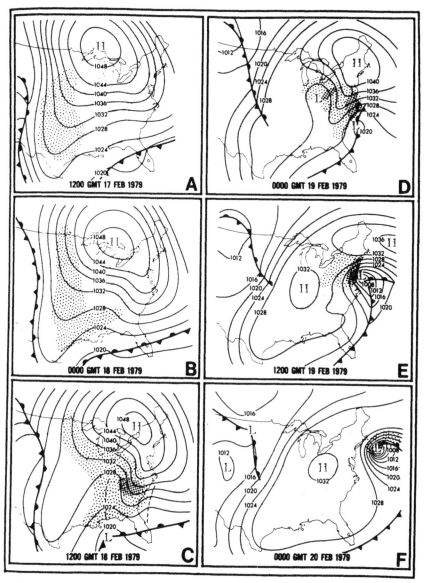

Fig. 5.11. Sea-level pressure (mb) and surface frontal analyses for (A) 1200 GMT February 17; (B) 0000 GMT February 18; (C) 1200 GMT February 18; (D) 0000 GMT February 19; (E) 1200 GMT February 19; and (F) 0000 GMT February 20, 1979. Shading represents precipitation; dark shading indicates moderate to heavy precipitation. Dashed lines in (C) denote inverted and coastal troughs. [After Uccellini *et al.* (1984).]

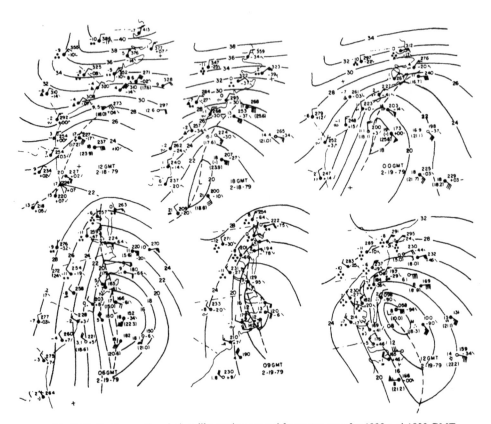

Fig. 5.12. Surface sectional plots illustrating coastal front structure for 1200 and 1800 GMT February 18 and 0000, 0600, 0900, and 1200 GMT February 19, 1979. Winds in meters per second [full (half) barb = 5 (2.5) m s^{-1}; pennant = 25 m s^{-1}] and temperatures in degrees Celsius. Surface isobars (mb) are given by solid lines; coastal front position illustrated by dashed lines. Open circles over ocean indicate buoys for which no weather information is presently available. Ocean temperatures (°C) are given in parentheses. [After Bosart (1981).]

approaching the North Carolina coast at 0600 GMT, with the coastal front now extending northward just west of Norfolk, Virginia, to the lower Chesapeake Bay. From 0900 to 1200 GMT the cyclone center passes west of Cape Hatteras and on into the western Atlantic as explosive deepening occurs. The coastal front reaches up into the northern Chesapeake Bay area at 0900 GMT and then retreats seaward after 1200 GMT in response to strong offshore cyclogenesis.

Another area having a high frequency of cyclogenesis is the western Gulf of Mexico. Saucier (1949) prepared a climatology of Texas–West Gulf cyclones for a 40-year period. An average of 9.7 cyclones per year developed across the region examined. The periodicity between cyclone maxima was 11 years. The majority of cyclones developed between 25°N and 30°N from 90°W to 99°W. Using 1972–1982 as a control period, an average of 10.4 winter cyclones developed each year over the Gulf of Mexico. Of these, 5.5 cyclones per year developed central pressures at or below 1010 mb (Johnson *et al.*, 1984). According to their study, from November 1982 through March 1983 a total of 26 surface cyclones affected the Gulf region. Five of these met the criteria established by Murty *et al.* (1983) for meteorological bombs, a pressure fall of 12 mb per 24 hr at 25°N. During that period the mean jet stream was about 5° farther south than normal over the Gulf of Mexico (see Fig. 5.13) (Quirox, 1983). This southward displacement of the mean polar jet stream was an important factor in the formation of the Gulf lows.

Two major patterns related to cyclogenesis in the northwestern Gulf of Mexico have been investigated by Johnson *et al.* (1984). They are the upper vortex and anticyclonic shear.

1. *Upper vortex pattern.* During the winter of 1982–1983 three of the five intense cyclones developed from a midtropospheric cyclone building toward

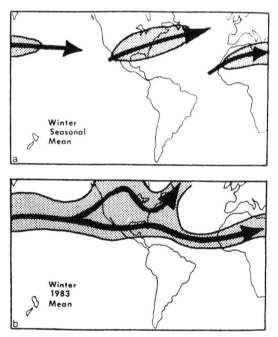

Fig. 5.13. The jet-stream pattern. (a) Location of the seasonal mean jet stream. (b) Location of the winter 1983 jet stream. [After Quirox (1983).]

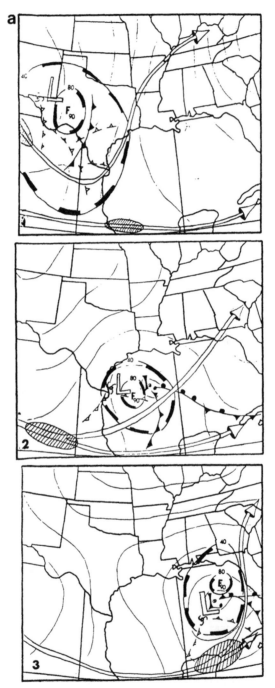

Fig. 5.14. (a) Synoptic composite charts for the Upper Vortex Pattern. (1) At 24 hr prior to explosive deepening. (2) Onset of rapid deepening. (3) Peak intensity 24 hr after onset of rapid deepening. [After Johnson *et al.* (1984).] (b) Symbols used in synoptic composite charts. [After Crisp (1979).]

	300 mb: Jetstream
	300 mb: Speed maximum
	500 mb: Cyclone
	500 mb: Shortwave trough
	500 mb: Vorticity maximum
	500 mb: Vorticity lobe
	500 mb: Height falls (m 12h^{-1})
	500 mb: Maximum height fall
	Surface: Cyclone
	Surface: Frontal boundary
	Surface: Isobars at 4 mb

Fig. 5.14. *(continued)*

the surface. As seen in Fig. 5.14a [for symbols, see Fig. 14b], a 500-mb cyclone moved out of the southwest United States into the northwest Gulf. As the accompanying 500-mb height falls moved into the Gulf, a surface low developed on the cyclonic shear side of the polar jet stream. Large, 500-mb height falls and the advection of positive vorticity (PVA) were sufficient to tap considerable latent and sensible heat sources from the lower troposphere over the Gulf. The combination led to rapid intensification of the surface cyclone during a 24-hr period. As the 500-mb height fall pattern shifted away from the surface low, the central pressure stabilized and gradually began to fill.

2. *Anticyclonic shear pattern.* Two of the five intense cyclones developed under conditions of anticyclonic shear from the polar jet stream (Fig. 5.15). A weak lee-side low developed near the Texas coast under the influence of weak diffluence and PVA. Low-level latent and sensible heat were tapped, and the surface cyclone began to deepen rapidly as (1) a moderate 500-mb height fall approached from the west, (2) PVA increased from the southwest, and (3) a wind speed maximum moved to a location placing the cyclone in its right-rear quadrant. The low reached maximum intensity and began filling as the 500-mb height fall pattern shifted northeastward away from the system.

Cyclogenesis in coastal regions is also common in the Yellow Sea between China and Korea, as well as over the East China Sea (Hanson and Long, 1985). An example is shown in Fig. 5.16. At 06Z on November 5, 1972, the only low-pressure system was over the China mainland, far from the Yellow Sea. It

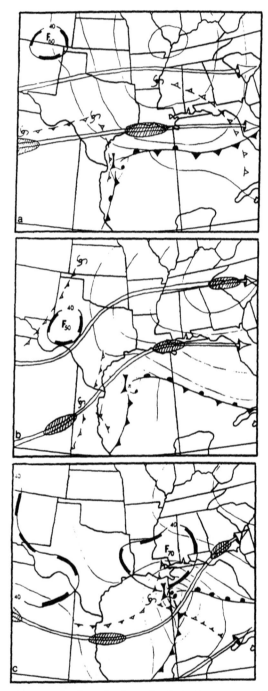

Fig. 5.15. Synoptic composite charts for the anticyclonic shear pattern. (a) At 24 hr prior to explosive deepening. (b) Onset of rapid deepening. (c) Peak intensity 24 hr after onset of rapid deepening. [After Johnson *et al.* (1984).]

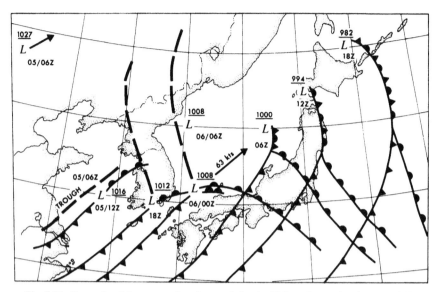

Fig. 5.16. Example of fronto- and cyclogenesis in the Yellow Sea region between 06Z on November 5 and 12Z on November 6, 1972.

propagated northeast and was not related to the system. A low-pressure trough developed over the Yellow Sea from Shanghai, China, to the northeast. This was the first hint of cyclogenesis.

By 12Z on November 5 a low-pressure center had formed. A warm front extended northeast and a cold front extended southwest from its center. Pressure was 1016 mb.

By 18Z the low (incipient cyclone) was moving eastward at 36 knots and deepened to 4 mb to a central pressure of 1012 mb. A pressure trough developed north through the Korean peninsula.

By 00Z on November 6 the cyclone had recurved to a northeast direction after crossing the southern tip of the Korea peninsula, moving at 28 knots. The central pressure deepened 5 mb to 1008 mb, with a trough extending north just off the east coast of Korea. The cyclone accelerated over the Sea of Japan (warm sector) to northeast at 63 knots, outrunning the trough.

By 06Z on November 6 the central pressure of the cyclone had deepened 8 mb to 1000 mb and started to occlude. Another low developed from the trough that was outrun by the cyclone. The cyclone continued to move northeast, but its speed had slowed to 38 knots over the Japanese Islands.

By 12Z on November 6 the cyclone's central pressure had deepened 6 mb to 994 mb, and moved northeast at 40 knots.

By 18Z on November 6 the cyclone was continuing northeast, and the central pressure had dropped 12 mb to 982 mb.

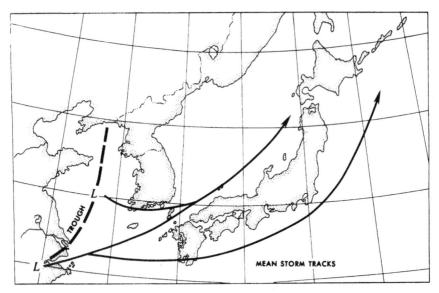

Fig. 5.17. Mean storm tracks at the surface in the region of the Yellow Sea, East China Sea, and Sea of Japan (see text for explanation).

The cyclone continued northeast, tracking south of Kamchatka Peninsula, and brought wind speeds of 30–45 knots through the straits and over the northwest Pacific, the pressure deepening to 954 mb.

The mean storm tracks at the surface in this region are shown in Fig. 5.17. This set of mean cyclone tracks has been synthesized from synoptic weather charts from the fall of 1972 and the fall of 1979. Spawning grounds are shallow-water regions near the center of the Yellow Sea and the western part of the East China Sea, where strong sea-surface temperature gradients (cool in the north and warm in the south) occur in arcuate patterns (concave southeast) in shallow water.

5.4 Tropical Storms

"Tropical storm" is a general term for a cyclone that originates over a tropical ocean (Fig. 5.18). At maturity, the tropical cyclone is one of the most intense and feared storms in the world; winds exceeding 90 m s^{-1} (175 knots or 200 miles per hour) have been measured, and rains are torrential (Huschke, 1959). Many books have been written on this subject (see, e.g., Riehl, 1979;

Fig..5.18. Examples of tropical storms in the eastern North Pacific on August 24, 1974. (Courtesy of U.S. National Oceanic and Atmospheric Administration.)

Simpson and Riehl, 1981; Anthes, 1982). Before the structure of tropical storms is discussed, some understanding of the tropical atmosphere may be helpful.

The intertropical convergence zone (ITCZ) or its variations such as "equatorial trough," "intertropical front," "monsoon and trade-wind troughs," and "buffer zone," is the axis, or a portion thereof, of the broad trade-wind current of the tropics. This axis is the dividing line between the southeast trades and the northeast trades (of the southern and northern hemispheres, respectively) (see Fig. 5.1). At one time it was held that this was a convergence line along its entire extent. It is now recognized that actual convergence occurs only along parts of this line (Huschke, 1959).

The ITCZ can be better discerned by studying the wind field near the surface, but away from frictional and other local effects. The gradient-level wind field is thus chosen. The gradient level is defined as the lowest level at which predominantly friction-free flow occurs. Over most tropical areas, this is taken to be 3000 ft (900 mb or 1 km) above mean sea level. Figures 5.19 through 5.22 show the resultant gradient-level streamlines and isotachs (lines of equal wind speed) for January and July conditions.

Atkinson (1971) provides the basis for the following brief discussion of the major features on gradient-level charts for midwinter and summer months. Note that monsoon troughs occur near the large continental areas owing to land–sea monsoonal effects and are characterized by a directional shear zone with westerlies on the equatorial side and easterlies on the poleward side. The trade-wind troughs occur primarily over oceanic areas of the north Atlantic and northeast and north-central Pacific and are characterized by the confluence of trade-wind flows from the northern and southern hemispheres. A buffer zone separates two wind systems of opposing directions, the easterly trades and the monsoon westerlies. The sense of rotation in the buffer zone varies from clockwise during the northern hemisphere (NH) summer to counterclockwise during the NH winter. The zone has time and space continuity and contains synoptic-scale cells.

5.4.1   January Characteristics

1. The northern hemisphere (NH) tropics are dominated by the oceanic subtropical ridges, which merge with the continental anticyclones.

2. Easterly flow south of this ridge system has a maximum speed near 10°N.

3. The trade-wind troughs and buffer zones are located between the equator and 5°N.

4. In the southern hemisphere (SH) tropics the monsoon trough is continuous from near the Fiji Islands westward across northern Australia and the south Indian Ocean into southern Africa and is the major producer of tropical cyclones in the southern oceans. Another important feature is the mean trough in the easterly trades extending southeastward from the low near 13°S, 170°E. This trough is associated with a distinct minimum in the isotach field, indicating a region of variable wind conditions associated with frequent synoptic disturbances. Similar isotach minima do not occur in the NH mean trade-wind flow. The region of this South Pacific trough is associated with a relative maximum of cloudiness and precipitation throughout the year.

5. The SH subtropical ridge is generally poleward of 30°S during January.

Fig. 5.19. Resultant gradient-level wind for January. [Adapted from Atkinson (1971).]

Fig. 5.20. Resultant gradient-level wind for January. [Adapted from Atkinson (1971).]

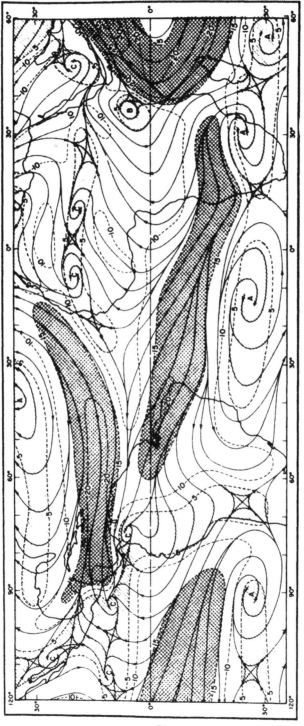

Fig. 5.21. Resultant gradient-level wind for July. [Adapted from Atkinson (1971).]

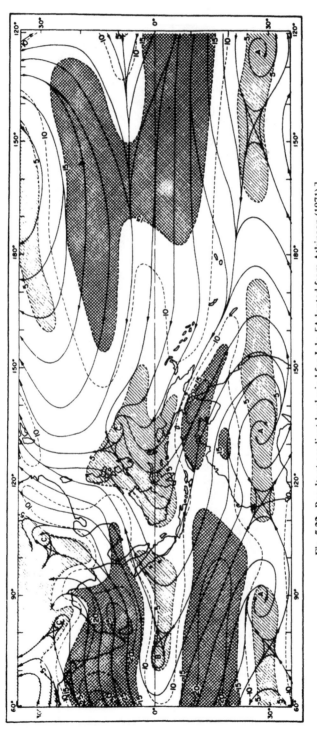

Fig. 5.22. Resultant gradient-level wind for July. [Adapted from Atkinson (1971).]

5.4.2   July Characteristics

1. The NH oceanic anticyclones have moved well northward and expanded. The trade-wind isotach maxima south of these anticyclones are located near 15–20°N.

2. The cyclone in the monsoon trough east of the Philippines is shown to indicate an area of active tropical storm formation during July.

3. Heat lows are apparent over Saudi Arabia and Mexico, and the monsoon trough over Africa has moved to near 20°N.

4. The northeast Pacific trade-wind trough of the cool season has now become a monsoon trough located near 10°N. This trough is an active producer of tropical storms during the NH summer and fall.

5. The near-equatorial buffer zone extends across almost half a hemisphere from 20°W to 140°E. Note that the mean axis of this zone crosses the equator several times. Two cells are embedded in the buffer zone, one south of India and one over Ethiopia.

6. The SH resultant gradient-level circulation has simplified considerably, the subtropical ridge and easterly trade-wind flow dominating the entire SH tropics.

7. A minor trough in the trade-wind easterlies is still evident in the South Pacific.

5.4.3   Easterly Waves

The easterly wave is a migratory wavelike disturbance of the tropical easterlies. It is a wave within the broad easterly current and moves from east to west, generally more slowly than the current in which it is embedded. Although best described in terms of its wavelike characteristics in the wind field, it also consists of a weak trough of low pressure. Easterly waves do not extend across the equatorial trough; parts of it could be described as ITCZ (Huschke, 1959). Note that the isobars usually display an inverted V shape when an easterly wave exists in the northern hemisphere (see Fig. 5.23).

According to Kotsch (1983), easterly waves are extremely important phenomena because of their relation to tropical cyclone, hurricane, and typhoon formation. On the average, easterly waves occur about every 15° of longitude during the summer season primarily and have an average length of 15–18° of latitude. They extend vertically in the atmosphere from the earth's surface to roughly 26,000 ft and travel from east to west at an average speed of 10–13 knots. Note in Fig. 5.23 that easterly waves are at right angles to the air flow and that the wave amplitude decreases with increasing latitude. Rather than being distinct air mass boundaries (such as fronts), easterly waves are zones of transition 30–100 miles wide in which the weather changes gradually

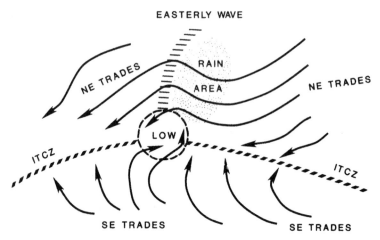

Fig. 5.23. A combination of easterly wave and the ITCZ may form a tropical depression. [Modified from Kotsch (1983).]

but very definitely. A vertical model of the normal type of easterly wave (Kotsch, 1983) shows that the average slope is in a ratio of 1/70 and that the easterly wave (EW) slopes upward from west to east. This, coupled with the fact that convergence of air flow occurs to the east of the EW and divergence to the west of the EW, results in the bad weather occurring behind—to the east of—the EW, as shown in Fig. 5.23.

5.4.4 Principal Storm Tracks

Major storm tracks in the tropics are illustrated in Fig. 5.24. Their sources and numbers generated during a 20-year period are shown in Fig. 5.25. Note that tropical storms follow a basic parabolic path (Fig. 5.24). According to Simpson and Riehl (1981), these cyclones are "steered" by the great anticyclone overlying tropical oceans. In the deep tropics they move westward in the easterlies and also tend to drift poleward. This drift sooner or later brings them to the limit of the tropics, where the east winds change to west winds. Accordingly, they "recurve."

As shown in Figs. 5.24 and 5.25, tropical cyclones occur in all warm-water oceans except the south Atlantic. This is because the ITCZ does not exist in the south Atlantic. Note that the southwestern part of the north Pacific has more tropical cyclones (typhoons) than any other place in the tropics. In this region the sea surface temperature (SST) is generally warmer than 28°C. In fact, this vast area is one of the warmest SST regions in the world's oceans.

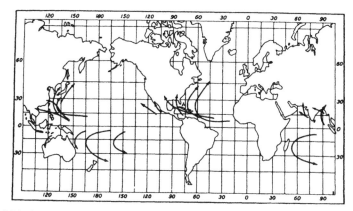

Fig. 5.24. Areas where tropical cyclones form, showing principal directions of paths. [After Dunn and Miller (1960).]

5.4.5 Formation

There are many papers on the subject of the formation of tropical storms. One of the simplest and most easily understood explanations is that of Kotsch (1983). The following discussion is based on that book.

One situation that is very conducive to the formation of a tropical cyclone (hurricane, typhoon, or other names for the same phenomenon) occurs when an easterly wave (EW) moves into juxtaposition with a northward "bulge" of the ITCZ (Fig. 5.23). In Fig. 5.23 the NE trade winds of the northern hemisphere have already developed a cyclonic pattern because of the presence

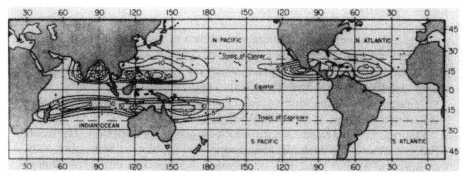

Fig. 5.25. Global sources of tropical cyclones and numbers generated during 20 tropical cyclone seasons (1958–1978). [After Simpson and Riehl (1981).]

Table 5.2

Conditions Favorable (or Necessary) for the Formation of Hurricanes and Typhoons[a]

---

1. Sea-surface temperature higher than 26°C (78.8°F).
2. Below-normal pressure in low latitudes (less than 1004 mb, or 29.95 inches) and above-normal pressure in higher latitudes.
3. An existing tropical disturbance of some sort at the earth's surface.
4. Movement of the disturbance at a speed less than 13 knots.
5. Easterly winds decreasing in speed with height, but extending upward to at least 30,000 ft.
6. Special dynamic conditions in the air flow near 40,000 ft.
7. Heavy rain or rainshowers in the area.

---

[a] After Kotsch (1983).

of the EW. As the SE trade winds of the southern hemisphere (where winds are deflected to the left) cross the equator and enter the northern hemisphere, they are deflected to the right and become SW winds, just about completing a closed cyclonic circulation. Hurricanes and typhoons derive their tremendous energy (and violence) from the latent heat of condensation (598 calories per gram) released into the atmosphere as the water vapor condenses. As long as the cyclonic center remains over warm water, the supply of energy is almost limitless. As more and more moist air spirals inward toward the low-pressure center to replace the heated and rapidly ascending air, more and more heat is released into the atmosphere and the wind circulation continues to increase. When the wind speeds exceed 64 knots (74 mph) we have, by definition, a hurricane or typhoon. Although we have no universally accepted, detailed theory of formation, most meteorologists agree that certain conditions are favorable (or necessary) for hurricane and typhoon formation. These conditions are contained in Table 5.2.

5.4.6   Structure

A general model for the hurricane in shown in Fig. 5.26. Note the strong low-level inflow to the core and corresponding outflow at high levels. At the center of a hurricane an "eye" exists, which exhibits calm wind and clear sky. In the northern hemisphere the primary energy cell is usually located on the right-hand side of the storm track. The spiral bands are associated with convective activities, as shown in the figure.

Figures 5.27–5.32 lend support to this general description. Hurricane Frederic struck the Alabama–Mississippi Gulf Coast in September 1979, causing widespread wind and storm-surge damage. The estimated damage total of $2.3 billion makes Frederic the costliest U.S. hurricane in history (Hebert, 1980). Frederic was monitored for 3 days—before, during, and after

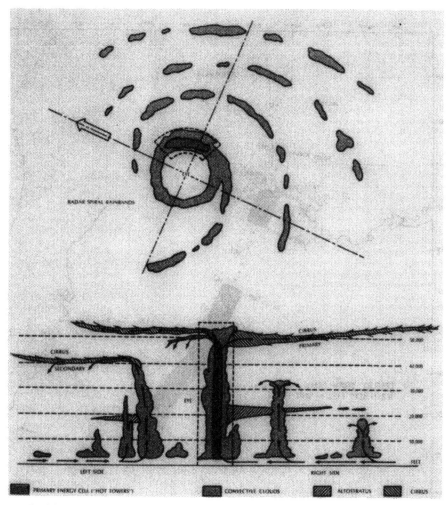

Fig. 5.26. A general model for the hurricane. The primary energy cell (called the convective chimney) is located within the area enclosed by the broken line. [From U.S. National Weather Service (1965).]

landfall—by seven National Oceanic and Atmospheric (NOAA) Research Facility Center scientific reconnaissance flights, numerous ship and buoy platforms in the open Gulf of Mexico, and many land stations. Frederic's track through the Gulf of Mexico and landfall in the moderately populated Pascagoula (Mississippi)–Mobile (Alabama) area was well documented (Powell, 1982). Figures 5.27–5.33 provide some of the characteristics of this hurricane.

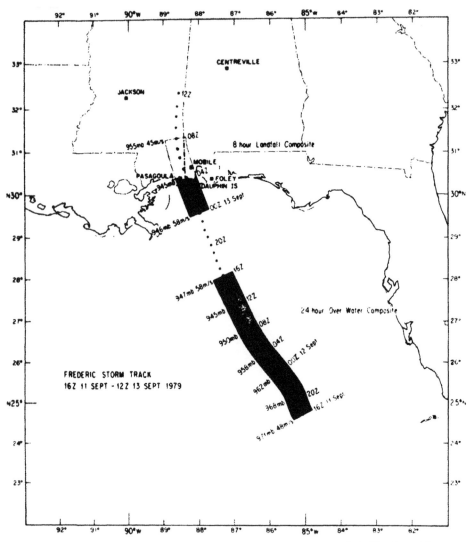

Fig. 5.27. Storm track and composite time periods for Hurricane Frederic. Hourly positions are noted in "Z" (GMT). Composite time periods are shaded; minimum pressures and maximum sustained wind speeds are included. [After Powell (1982).]

Fig. 5.28. The well-defined eye of Hurricane Frederic as it approaches the Alabama coastline on September 12, 1979. [After Hebert (1980).]

The track of Frederic, as recorded by Powell (1982), including maximum flight-level sustained wind speeds and minimum sea level pressures, is shown in Fig. 5.27. Frederic reintensified into a hurricane at 1200 GMT on September 10, 1979, over the western end of Cuba and moved north-northwest during the next 3 days, making landfall just east of the Alabama–Mississippi border at 0400 GMT on September 13. A photo of Frederic taken by a weather satellite is shown in Fig. 5.28. This picture resembles the drawing of a typical hurricane on the upper panel of Fig. 5.26. A "synoptic" representation of the hurricane in the central Gulf of Mexico, with no land influences, is shown in Fig. 5.29. The circulation around the eye is very evident. Note that from 16Z on September 11 to 16Z on September 12 (Fig. 5.27) the storm motion was approximately 5 m s^{-1} and pressure dropped from 971 to 945 mb, for an average drop of 1 mb per hour, while the maximum sustained flight-level wind speed increased from 48 to 58 m s^{-1}.

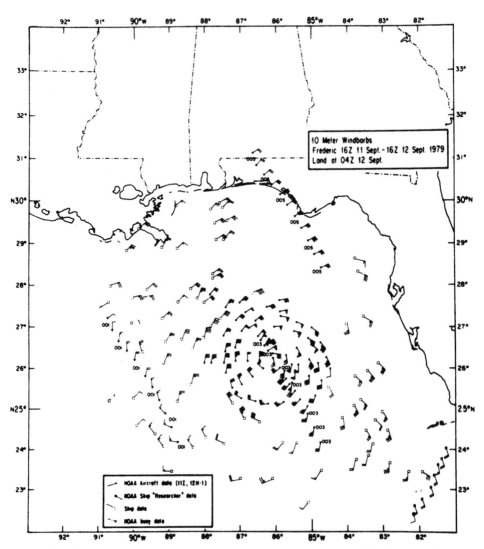

Fig. 5.29. Data coverage for the overwater composite. Numbers refer to NDBO buoys. Wind barbs at 10-m level correspond to conventional plotting, with speeds in knots. [After Powell (1982).]

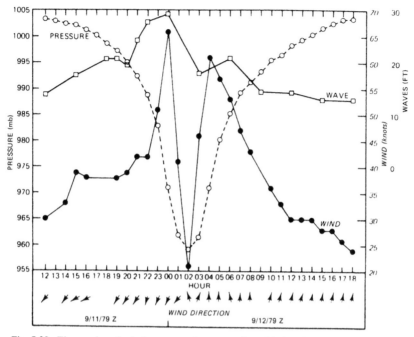

Fig. 5.30. Time series of wind, wave, and pressure from National Weather Service (NWS) Buoy 42003, located at 26°N, 86°W, which was directly in the path of Hurricane Frederic, September 11–12, 1979. [After De Angelis (1980).]

According to De Angelis (1980), when Hurricane Frederic was moving to the northwest in the Gulf of Mexico on September 11 and 12, Environment Buoy 42003, at 26°N, 86°W, was directly in its path (see Fig. 5.27). Figure 5.30 is a graph of regular synoptic reports of pressure, wind, and wave height for the dates and hours shown. Waves were recorded irregularly during the period. The lowest pressure recorded by the buoy was 959.3 mb at 02Z on September 12. The highest sustained wind recorded was 66 knots, and the highest wave was nearly 30 feet, both recorded at 00Z on September 12. The pattern shown is a classic hurricane passage, with pressures dropping precipitously, then rising in nearly a mirror image. The northeasterly winds built to their highest speed in advance of the center (as the movement of the storm reinforced their strength), dropped to moderate breezes in the eye, then veered abruptly to the south with renewed, near-hurricane force. The waves built in advance of the storm, then began a gradual, irregular decay.

The circulation pattern of Frederic near the time of landfall is shown in Fig. 5.31. Its spiral bands are clearly evident from the radarscope shown in Fig. 5.32. Note that Fig. 5.32 further supports the general hurricane model, as shown on the upper panel of Fig. 5.26.

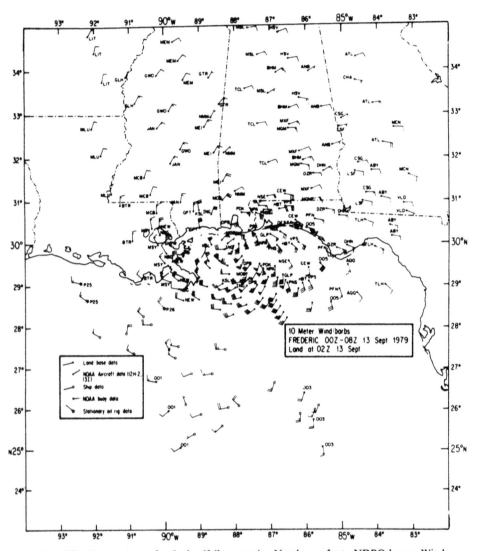

Fig. 5.31. Data coverage for the landfall composite. Numbers refer to NDBO buoys. Wind barbs at 10-m level correspond to conventional plotting, with speeds in knots. [After Powell (1982).]

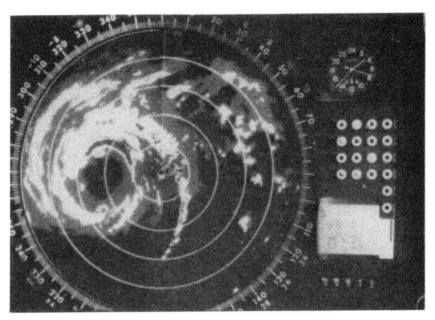

Fig. 5.32. Pensacola, Florida, radar shows the eye of Frederic in Mobile Bay early on September 13, 1979. [After Hebert (1980).]

As mentioned in Section 4.3.3, the wind speed of hurricanes approximately obeys the cyclostrophic flow. The equation states that

$$\frac{C^2}{r} = \frac{1}{\rho}\left(\frac{\partial P}{\partial r}\right) \tag{5.10}$$

where C is the wind speed, r is the radius of the hurricane, ρ is the air density, and $\partial P/\partial r$ is the radial pressure gradient.

Since Hurricane Frederic was well documented, including the radial profiles of wind speed and surface pressure (see Fig. 5.33), it would be interesting to apply Eq. (5.10) to this case. For September 11 (see 11I in the figure),

$$r = 150 \quad \text{km}$$

$$\frac{\partial P}{\partial r} \simeq \frac{\Delta P}{\Delta r} = \frac{(998 - 962)}{150} \quad \frac{\text{mb}}{\text{km}}$$

$$\rho \simeq 1 \quad \text{kg m}^{-3}$$

Since 1 mb = 10^2 Pa = 10^2 kg m^{-1} s^{-2}

$$C = 60 \quad \text{m s}^{-1}$$

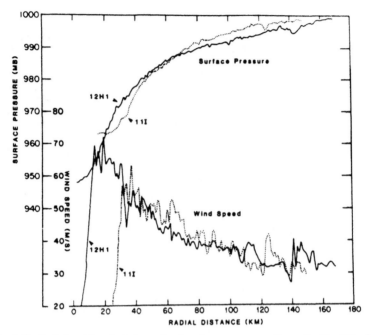

Fig. 5.33. Radial profiles of wind speed and surface pressure on the north side of Frederic at the 500-m level for September 11 (11 I) and September 12 (12 H1), 1979, flights. [After Powell (1982).]

Similarly, for September 12 (see 12 H1 in Fig. 5.33),

$$r = 170 \quad \text{km}$$

$$\frac{\partial P}{\partial r} = \frac{\Delta P}{\Delta r}$$

$$= \frac{(998 - 948) \quad \text{mb}}{170 \quad \text{mb}}$$

$$\rho \simeq 1 \quad \text{kg m}^{-3}$$

we get

$$C = 70 \quad \text{m s}^{-1}$$

Both values of C are in excellent agreement with their respective maximum 30-s average flight-level (at 500 m) wind speed as shown in Fig. 5.33. Thus, we conclude that the cyclostrophic equation is a useful formula for estimating the approximate speed of a hurricane.

5.5 Some Aspects of Weather Analysis and Forecasting

In synoptic meteorology, weather analysis is a detailed study of the state of the atmosphere based on actual observations, usually including a separation of the entity into its component patterns and involving the drawing of families of isopleths (lines of equal or constant value of a given quantity) for various elements. Thus analysis of synoptic charts (or weather maps) may consist of, for example, the drawing and interpretation of patterns of wind, pressure, pressure change, temperature, humidity, clouds, and hydrometeors (precipitation), all based on actual simultaneous observations (Huschke, 1959).

Weather forecasting requires weather chart analyses from both series of surface weather maps and upper-air flow charts. In order to do this, the important aspects of synoptic meteorology are broken down into the following four steps (Williams *et al.*, 1968):

1. Observing and reporting the weather.
2. Collecting and displaying the data.
3. Analyzing the information.
4. Interpreting and forecasting the weather.

Weather information is first recorded by weather observers in many places at regular times each day and then sent by fast communication systems to the national meteorological center, which prepares the weather maps.

For land stations the coded messages are as follows:

$$\text{iii } Nddff\ VVwwW\ PPPTT\ N_hC_LhC_MC_H\ T_dT_d\text{app } 7RRR_ts$$

where

iii	Station identification number
N	Sky coverage (total amount)
dd	True direction *from* which wind is blowing
ff	Wind speed in knots or miles per hour
VV	Visibility in miles
ww	Present weather
W	Past weather
PPP	Barometric pressure in millibars reduced to sea level
TT	Current air temperature in Fahrenheit (used in North American weather maps)
N_h	Fraction of sky covered by low or middle clouds
C_L	Low clouds or clouds of vertical development
h	Height in feet of the base of the lowest clouds
C_M	Middle clouds
C_H	High clouds

T_dT_d Dew-point temperature in Fahrenheit (used in North American weather maps)

a   Pressure tendency

pp   Pressure change (in millibars) in preceding 3 hr ($+28 = +2.8$ mb)

RR   Amount of precipitation ($45 = 0.45$ inch)

R_t Time precipitation began or ended

These messages are plotted as Land-Station Plotting model:

$$
\begin{array}{c}
/ \\
ff \\
\diagdown \quad C_H \\
TTdd \; C_M \; PPP \\
VVww \; \textcircled{N} \; \pm \; ppa \\
T_dT_d \; C_L \, N_h \; W R_t \\
h \quad RR
\end{array}
$$

The U.S. ship plotting model is

U.S. Ship Model

$$
\begin{array}{c}
/ \\
ff \\
\diagdown \\
\qquad C_h \; D_s \; V_s \\
dd \; Cm \; PPP \\
TT \; \diagdown \\
VVww \textcircled{N} \pm ppa \\
T_dT_d \; C_L \; N_h \; W \\
(T_s T_s) h (d_w d_w P_w H_w)
\end{array}
$$

Their messages are as follows:

$99L_aL_aL_aQ_cL_oL_oL_oL_o$ $YYGGi_w$ $Nddff$ $VVwwW$ $PPPTT$ N_h $C_LhC_MC_H$

$D_sV_s$app $OT_sT_sT_dT_d$ $3P_wP_wH_wH_w$

A definite order of decoding has been established, as in the land station code, to eliminate confusion between the decoder and plotter on the order of entries.

The first elements to be checked by the decoder are the day of the month (YY) and the time (GG), to be sure that the report is consistent with the date and time the chart is being plotted.

After determining the quadrant (Q_c) of the globe, the first verbally decoded element is latitude ($L_aL_aL_a$), and this is always decoded as "north" or "south."

The second item decoded is the longitude ($L_oL_oL_oL_o$), and this is always decoded as "west" or "east."

When the plotter has located the position indicated, he should draw a $\frac{1}{8}$-inch circle over it.

The decoder then proceeds to decode in the following manner:

dd	Wind direction
ff	Wind velocity
N	Total sky coverage
PPP	Sea-level pressure
TT	Temperature
ww	Present weather
VV	Visibility (from ship's visibility code table)
T_dT_d	Dewpoint temperature
T_sT_s	Seawater temperature
app	Pressure tendency (pressure change, as recorded on shipboard, is affected by the movement of the ship; therefore, the ship's movement should be shown whenever pressure change is entered)
D_s	Ship's average direction
V_s	Ship's average speed
C_m	Middle cloud type
C_h	High cloud type
C_L	Low cloud type
N_h	Amount of low clouds
h	Height of lowest cloud
$P_wP_wH_wH_w$	Wave period and height

Note that the coded digital message for the above symbols is explained in various tables, which may be found in many elementary meteorology textbooks or exercise manuals. More complete explanations are available in the "Manual for Synoptic Code" from the Superintendent of Documents, U.S. Government Printing Office, Washington, D.C. Readers may also ask nearby National Weather Service stations for proper explanation.

After the data of each station are collected and decoded, they are plotted on the surface chart for that station. Examples are shown in Fig. 5.34. Then, isopleths of certain elements such as the pressure (isobars) can be drawn as shown in Fig. 5.35. Note that Fig. 5.34 is a segment of unanalyzed chart, whereas Fig. 5.35 is the analyzed chart (in this case, the isobars and the cold front). Note also that the pressure should be reduced to sea level according to the procedures provided by the World Meteorological Organization (WMO). To decode the pressure in millibars, one should be aware of the following steps:

1. Place a decimal point to the left of the last number. For example, for the topmost two stations on the upper right-hand corner in Fig. 5.34, 000 becomes 00.0 and 988 becomes 98.8.

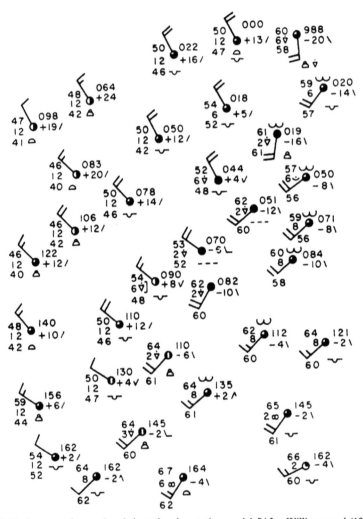

Fig. 5.34. Segment of unanalyzed chart showing station model. [After Williams *et al.* (1968).]

2. Next, place either a 9 or 10 in front (to the left) of the first digit. Thus, 00.0 becomes either 900.0 or 1000.0 and 98.8 becomes 998.8 or 1098.8.

3. To determine whether to place a 9 or 10 in front of the first digit: (a) If the number falls between 56.0 and 99.9, place a 9 before the first digit. (b) If the number falls between 00.0 and 55.9, place a 10 before the first digit. Thus, 00.0 should be 1000.0 and 98.8 should be 998.8 mb. These two stations are then incorporated into the determination of the pressure field (in this case, a low-pressure center) in that region as analyzed in the top part of Fig. 5.35.

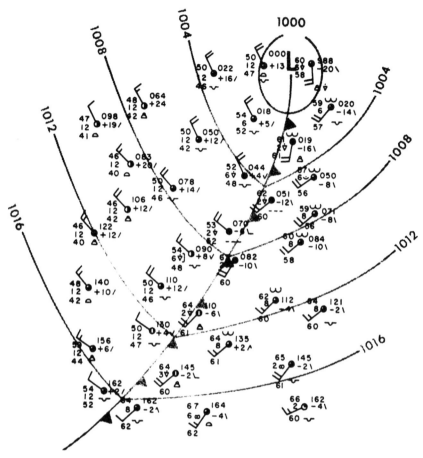

Fig. 5.35. Segment of analyzed chart showing station models. [After Williams *et al.* (1968).]

Daily weather maps of various types are available from a variety of sources: the U.S. National Weather Service, most daily newspapers, private weather consultants, industrial meteorological organizations, and so forth.

Figure 5.36 includes examples of NOAA Weather Service daily weather maps, which are available at a nominal cost from the Climate Analysis Center, World Weather Building, Washington, D.C. The surface weather map presents station data and the analysis for 0700 EST. The tracks of well-defined low-pressure areas are indicated by chains of arrows (see, e.g., along the eastern seaboard of the United States on this map). The locations of these centers 6, 12, and 18 hr preceding map time are indicated by small white crosses in black squares. Areas of precipitation are indicated by shading. The

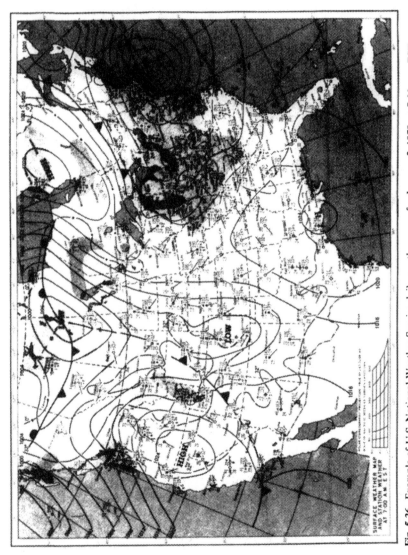

Fig. 5.36. Example of U.S. National Weather Service daily weather maps for January 2, 1978, at 7:00 a.m. EST.

weather reports printed on these maps are only a small fraction of those that are included in the operational weather maps and on which analyses are based. Occasional apparent discrepancies between the printed station data and the analyses result from those station reports that cannot be included in the published maps because of lack of space. The symbols of fronts, pressure systems, air masses, and so forth are the same as discussed in previous chapters. Note how the isobars are V-shaped (they "kink") at the fronts, as previously discussed.

The 500-Millibar Chart presents the height contours and isotherms of the 500-mb surface at 0700 EST. The 500-mb heights usually range from 4800 to 5900 m. The height contours are shown as continuous lines and are labeled in dekameters above sea level. The isotherms are shown as dashed lines and are labeled in degrees centigrade. The arrows show the wind direction and speed at the 500-mb level.

The Highest and Lowest Temperature Chart shows the maximum temperature for the 12-hr period ending at 7:00 p.m. EST of the previous day and the minimum temperature for the 12-hr period ending at 7:00 a.m. EST. The names of the reporting points' are shown on the Surface Weather Map. The maximum temperature is plotted above the station location, and the minimum temperature is plotted below this point.

The Precipitation Areas and Amounts Chart shows areas (shaded) that had precipitation during the 24 hr ending at 7:00 a.m. EST, with amounts to the nearest hundredth of an inch. Incomplete totals are underlined. A "T" indicates a trace of precipitation. Dashed lines, in season, show the depth of snow on the ground in inches at 7:00 a.m. EST.

Weather forecasts by professional meteorologists are prepared on the basis of many scientific principles, the assistance of high-speed computers as well as weather radars and satellites, and many years of experience. One of the easier ways to determine the future movement of pressure systems is to do a freehand extrapolation of their paths. This method has been appropriately named the "path method" by Professor S. Petterssen. For this method, a series of weather maps (rather than a single weather map) is required. Figure 5.36 illustrates one such path along the eastern seaboard of the United States as indicated by the center of the storm track. Basically, the "path method" of forecasting involves the extrapolation of established trends into the future—whether they be speed changes, intensity changes, directional-movement changes, some of them, or all of them.

In weather forecasting, upper-air charts must be incorporated because the direction and speed of movement (and other factors) of surface pressure systems and fronts are influenced to a great degree by the air currents (and other conditions) at higher levels in the atmosphere (see, e.g., Kotsch, 1983). If the speed of upper-air flow is slow and the height contours have large

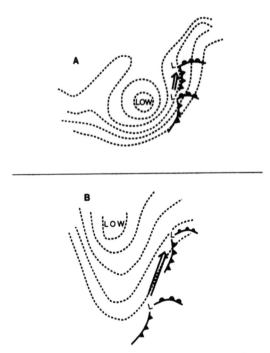

Fig. 5.37. Effect of upper-air pressure distributions (dotted curves) on the movement of surface frontal systems (solid curves) (see text for explanation).

curvature, the movement of surface frontal systems is also slow, as shown in Fig. 5.37A. This situation is illustrated by comparison between Figs. 5.36 and 5.38 for a time interval of 24 hr. On the other hand, if the speed of upper-air flow is fast and the height contours have small or no curvature, the movement of surface frontal systems is also fast, as shown in Fig. 5.36B. This effect can be seen by comparison between Figs. 5.39 and 5.40 for a time interval of 24 hr. From this, it is readily apparent that some knowledge and understanding of upper-air flow patterns is essential for accurate surface weather forecasting.

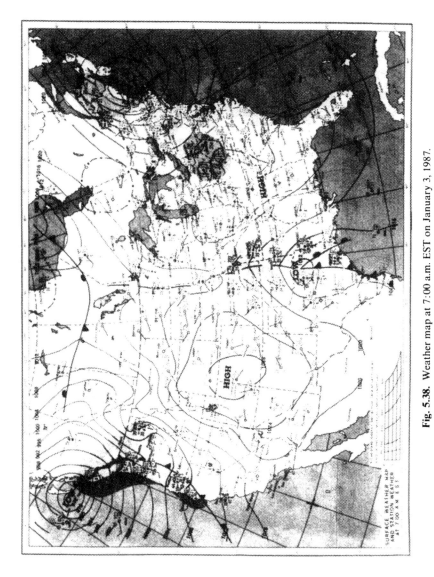

Fig. 5.38. Weather map at 7:00 a.m. EST on January 3, 1987.

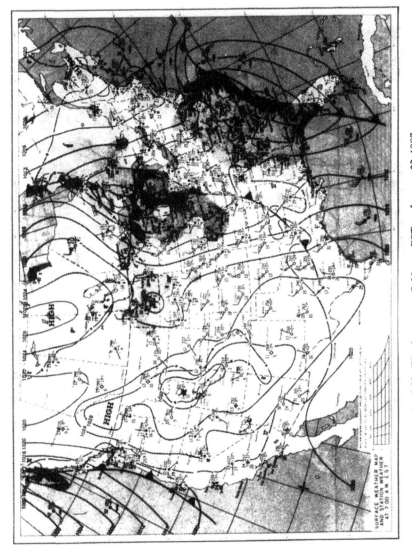

Fig. 5.39. Weather map at 7:00 a.m. EST on January 22, 1987.

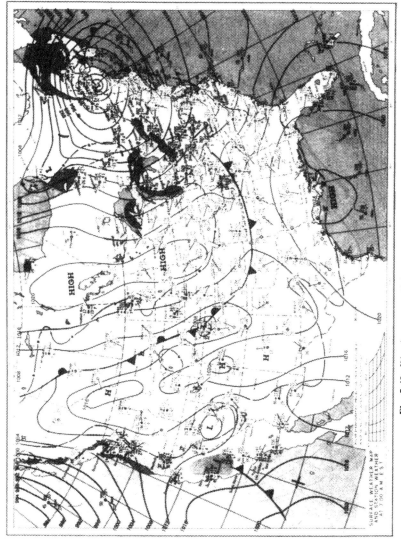

Fig. 5.40. Weather map at 7:00 a.m. EST on January 23, 1987.

Chapter 6 | Atmospheric Boundary Layers and Air–Sea Interaction

The atmospheric or planetary boundary layer (PBL) is the frictionally dominated region between the earth's surface and the geostrophic (frictionless or synoptic) flow level. As shown in Fig. 1.2, the PBL consists of the surface layer and the transition (or Ekman) layer. In order to understand momentum, heat, moisture, and aerosol (particles) transfer at the air–land and air–sea interfaces, study of the atmospheric surface boundary layer is very important in meteorology and physical oceanography. This chapter is devoted to the investigation of the PBL and smaller-scale air–coast and air–sea interactions.

Extensive studies and comprehensive summaries related to the PBL can be found in, for example, Munn (1966), Plate (1971), Haugen (1973), Wyngaard (1978), McBean (1979), and Panofsky and Dutton (1984).

6.1 The Surface Boundary Layer

The surface boundary layer (SBL), or simply the surface layer, is the lowest part of the PBL. Under conditions of horizontal homogeneity (e.g., over open oceans or broad, flat prairies) and quasi-steady state (e.g., time changes are so small as to be dynamically negligible), the following approximations are usually applied to the SBL (see Wyngaard, 1973; McBean, 1979; Panofsky and Dutton, 1984):

1. The rotation of the earth, that is, the Coriolis effect, is probably unimportant in the SBL.
2. The SBL occupies the lowest 10% of the PBL.
3. Experiments have shown that the vertical variation in stress and heat flux in the SBL is within 10% and thus the "constant" flux layer is also named for the SBL.

4. From (3) it also follows that in the SBL the wind direction does not change appreciably with height. Thus, the mean wind is described by $\bar{U}$ only.

5. The variation of mean variables with height Z, is controlled primarily by three parameters: the surface stress, the vertical heat flux at the surface, and the terrain roughness.

6. Transport of atmospheric properties by turbulent diffusion (eddies) is much more important than transport by molecular diffusion.

Based on the above conditions and in analogy to Eq. (4.42), vertical fluxes of momentum, heat, and moisture in the SBL can be written (see, e.g., Busch, 1973) so that

$$\text{Wind stress } \tau = -\rho_0\overline{u'w'} = \rho U_{*0}^2 \tag{6.1}$$

$$\text{Heat flux } H = C_p\rho_0\overline{\theta'w'} = -C_p\rho_0 U_{*0}T_* \tag{6.2}$$

$$\text{Moisture flux } E = \rho_0\overline{q'w'} = -\rho U_{*0}q_* \tag{6.3}$$

define the vertically invariant scaling parameters U_{*0}, T_*, and q_*, where u' and w' are the fluctuation of the horizontal wind speed U and vertical wind speed w, respectively (cf. Sections 4.4.2 and 4.4.3), θ' and q' are the fluctuations in potential temperature θ and specific humidity q, respectively, the overbars are the means, ρ_0 is the air density of reference state, C_p is the specific heat at constant pressure, and U_{*0}, T_*, and q_* are the friction (or shear) velocity of reference state, the characteristic temperature, and the characteristic specific humidity, respectively. The parameters U_{*0}, T_*, and q_* are the characteristic scales based on Monin–Obukhov (1954) similarity theory.

By applying this scaling to the gradients of the mean profiles such as Eq. (4.42), we have

$$\frac{\kappa Z}{U_{*0}}\left(\frac{\partial \bar{U}}{\partial Z}\right) = \phi_m\left(\frac{Z}{L}\right) \tag{6.4}$$

$$\frac{\kappa Z}{T_*}\left(\frac{\partial \bar{\theta}}{\partial Z}\right) = \phi_h\left(\frac{Z}{L}\right) \tag{6.5}$$

$$\frac{\kappa Z}{q_*}\left(\frac{\partial \bar{q}}{\partial Z}\right) = \phi_e\left(\frac{Z}{L}\right) \tag{6.6}$$

where κ is the von Karman constant ($\simeq 0.4$; see, e.g., Högström, 1985), defined such that $\phi_m(Z/L) = \phi_m(0) = 1$, m, h, e represent momentum, heat, and evaporation, respectively and

$$L = -\frac{U_*^3 T_{v0}}{g\kappa\overline{\theta_v'w'}} \tag{6.7}$$

is the Monin–Obukhov stability length, T_{v0} is the virtual temperature of the

Table 6.1

Qualitative Interpretation of Z/L^a

Z/L	Word description
Strongly negative	Heat convection dominant
Negative but small	Mechanical turbulence dominant
Zero	Purely mechanical turbulence
Slightly positive	Mechanical turbulence slightly damped by temperature stratification
Strongly positive	Mechanical turbulence severely reduced by temperature stratification

^{a} After Panofsky and Dutton (1984).

reference state, and θ_v' is the turbulent fluctuation of virtual temperature $(= T_v' - \overline{T_v'})$. The term T_v' is the virtual perturbation temperature $(= T_v - T_{v0})$, in which T_v is the virtual temperature and T_{v0} is the virtual temperature of reference state.

It can be seen that parameter

$$\frac{Z}{L} = -\frac{g\kappa Z\overline{\theta_v'w'}}{U_*^3 T_{v0}} \tag{6.8}$$

According to Eq. (6.8) and Panofsky and Dutton (1984), we can consider the ratio $-Z/L$ to represent the relative importance of heat convection (i.e., the buoyancy term $\overline{\theta_v'w'}$), shown in the numerator, and mechanical turbulence (i.e., the wind shear U_*), as shown in the denominator. The general properties of Z/L are summarized in Table 6.1.

Note that during the day the ground is warmer than the air at higher elevations. Therefore, $\overline{\theta_v'w'}$ is positive and Z/L is negative. Conversely, at night the ground is colder and the $\overline{\theta_v'w'}$ is negative (i.e., the heat is transferred from higher elevation to the ground). Thus Z/L is positive. Near sunrise and sunset $\overline{\theta_v'w'} \to 0$; thus from Eq. (6.7) $L \to \infty$ and from Eq. (6.8) $Z/L \to 0$ under these adiabatic or near-neutral conditions (i.e., $T_v = T_{0v}$ and $T_v' = \overline{T_v'}$ or $\theta_v' = 0$). However, mechanical turbulence caused by wind shear may be dominant under these conditions, as summarized in Table 6.1.

In Chapter 4 we introduced a similar stability parameter, namely, the Richardson number R_i, as defined in Eq. (4.46), which shows also the ratio of buoyancy to wind shear. The relationship between Z/L and R_i is thus obvious. Experiments have shown that (see Panofsky and Dutton, 1984) in unstable air (mainly during the day)

$$R_i = Z/L \tag{6.9}$$

and in stable air (mainly at night)

$$Z/L = R_i/(1 - 5R_i) \tag{6.10}$$

Returning to Eqs. (6.4) and (6.5), most atmospheric data are well represented (see, e.g., Busch, 1973; Panofsky and Dutton, 1984) by the following empirical formulations.

For unstable $(Z/L < 0)$ conditions,

$$\phi_m(Z/L) = (1 - 16Z/L)^{-1/4}$$
$$\phi_h(Z/L) = \phi_q(Z/L) = (1 - 16Z/L)^{-1/2} \tag{6.11}$$

For neutral $(Z/L = 0)$ conditions,

$$\phi_m(Z/L) = \phi_h(Z/L) = \phi_q(Z/L) = 1 \tag{6.12}$$

For stable $(Z/L > 0)$ conditions,

$$\phi_m(Z/L) = \phi_h(Z/L) = \phi_q(Z/L) = 1 + 5Z/L \tag{6.13}$$

We are now in a position to discuss various wind profiles in the surface layer.

6.2 The Logarithmic Wind Profile

It is very important to understand the vertical distribution of the wind or wind shear, $\partial U/\partial Z$, because it is related to the momentum transfer through Eqs. (6.1) and (6.4). The momentum flux in turn is the driving force to generate wind-drift ocean currents, for example. As discussed previously, under adiabatic or near-neutral conditions $Z/L \to 0$; then $\phi_m(Z/L) = \phi_m(0) = 1$ and Eq. (6.4) becomes

$$\frac{\kappa Z}{U_{*0}} \frac{\partial \bar{U}}{\partial Z} = 1$$

or

$$\frac{\partial \bar{U}}{\partial Z} = \frac{U_*}{\kappa Z}$$

According to Businger (1973), this expression indicates an infinite shear at the surface $(Z = 0)$, which of course is unrealistic. In order to keep the shear finite, a surface roughness Z_0 is introduced such that

$$\frac{\partial \bar{U}}{\partial Z} = \frac{U_*}{\kappa(Z + Z_0)}$$

Integration yields

$$\bar{U}_z = \frac{U_*}{\kappa} \ln\left(\frac{Z + Z_0}{Z_0}\right) \tag{6.14}$$

where $\bar{U}_z$ is the mean wind speed at height Z. Equation (6.14) is the well-known logarithmic wind profile. A schematic is shown in Fig. 6.1.

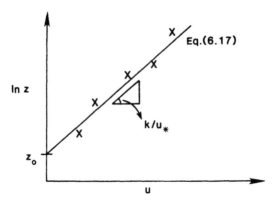

Fig. 6.1. A representation for Eq. (6.16) or Eq. (6.17).

From the values of Z_0 as shown in Panofsky and Dutton (1984, p. 123), we have $Z \gg Z_0$, and Eq. (6.14) becomes

$$\bar{U}_z = \frac{U_*}{\kappa} \ln \frac{Z}{Z_0} \tag{6.15}$$

or

$$\kappa \bar{U}_z / U_* = \ln Z - \ln Z_0$$

or

$$\ln Z = \ln Z_0 + (\kappa/U_*)\bar{U}_z \tag{6.16}$$

This equation has a least-square linear regression form such that (cf. Fig. 6.1)

$$Y = a_0 + a_1 X \tag{6.17}$$

where $Y = \ln Z$ and

$$a_0 = \ln Z_0 \quad \text{or} \quad Z_0 = e^{a_0} \tag{6.18}$$

$$a_1 = \kappa/U_* \quad \text{or} \quad U_* = \kappa/a_1 \tag{6.19}$$

and $X = \bar{U}_z$.

Examples of the logarithmic wind profile [Eq. (6.15)] on a flat, open coast are shown in Fig. 6.2. From these profiles, values of Z_0 and U_* can be obtained by Eqs. (6.18) and (6.19), respectively. A summary of these values in coastal environments is shown in Fig. 6.3. Note that for a given Z_0 and a preassigned reference height such as $Z = 2$ m Eq. (6.15) can be used to compute the relationship between U_* and U_{2m}. These relationships are also given in Fig. 6.3.

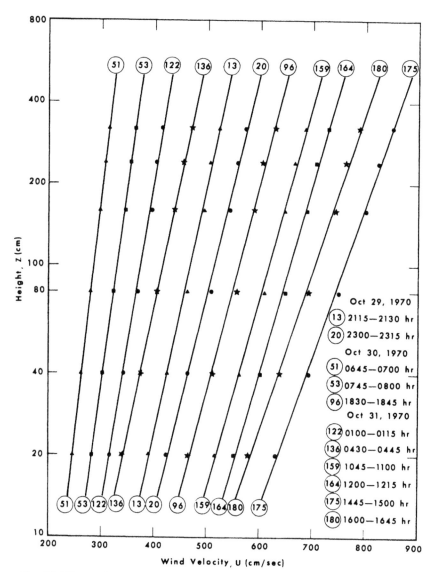

Fig. 6.2. Fifteen-minute mean wind speed as a function of height (on a logarithmic scale) on the mud flat of the tidal stream of Estero de Data, Gulf of Quayaquil, Ecuador. Circled numbers are the experimental sequence numbers. Symbols represent measuring points. [After Hsu (1972b). Copyright by D. Reidel Publishing Co. Reprinted by permission.]

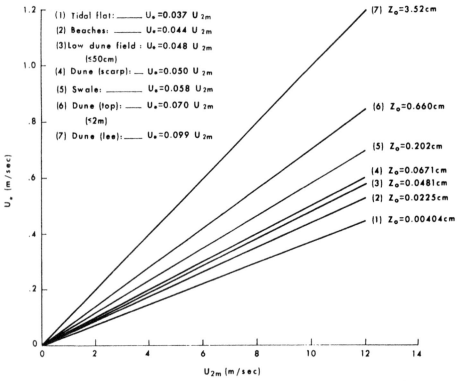

Fig. 6.3. Summary of the aerodynamic roughness length Z_0 and the relationship between shear velocity U_* and the wind velocity at 2 m height (U_{2m}) measured in various coastal environments for eolian sand transport estimation. Note that (1) was obtained from Ecuador, (2) is a synthesis of six beaches (Barbados, Ecuador, Texas, Brazil, Florida, and the Alaskan Arctic), (3) is from Texas, and (4)–(7) are all from Brazil. [After Hsu (1977).]

6.3   The Nonadiabatic Wind Profile

The nonadiabatic (or diabatic) process is defined as thermodynamic change of state of a system in which there is transfer of heat across the boundaries of the system (Huschke, 1959). As mentioned previously, during the day the temperature decreases with height, rapidly in the lower layers and more slowly at greater heights. At night, temperature increases with height—that is, the inversion case.

Figures 6.4 and 6.5 show the hourly variation of the temperature profile between 1.8 and 90.9 m for May 20 and 21, 1969, respectively. For reference, times of sunrise and sunset for these two days were approximately 0504 and

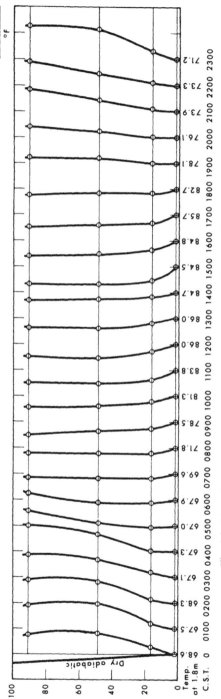

Fig. 6.4. Diurnal variation of temperature between 1.8 and 90.9 m at Eglin Air Force Base, near Fort Walton Beach, Florida, May 20, 1969. [After Hsu (1973).]

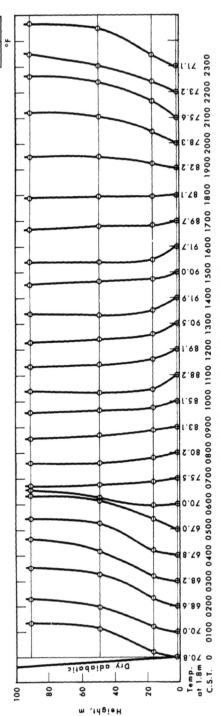

Fig. 6.5. Same as Fig. 6.4 for May 21, 1969.

1849 CST, respectively. Since the lapse rate γ of the actual temperature T is defined as the rate of T decrease with height Z (i.e., $\gamma = -dT/dZ$) and since the dry adiabatic lapse rate γ_a is $0.98°C/100$ m (or $5.4°F/1000$ ft), the stability criteria for the unsaturated air may be written as (e.g., Haltiner and Martin, 1957)

$$\gamma > \gamma_a \qquad \text{unstable}$$

$$\gamma = \gamma_a \qquad \text{neutral}$$

$$\gamma < \gamma_a \qquad \text{stable}$$

Comparing the hourly temperature profiles with the dry adiabatic lapse rate shown on the left of these figures, we see that, in general, conditions are stable at night and unstable during the day over the coastal zone.

The corresponding diurnal variations of wind speed between 3.6 and 90.9 m for May 20 and 21, 1969, are shown in Figs. 6.6 and 6.7, respectively. Note that the wind profiles are plotted on semilogarithmic paper so that a straight line may be fitted for the neutral stability. For the sake of clarity, profiles are plotted for every 3 hr instead of each hour. Figures 6.6 and 6.7 clearly show that the wind profile is concave either upward (during the day) or downward (at night) except around midmorning (0900 LST), when the transitional period

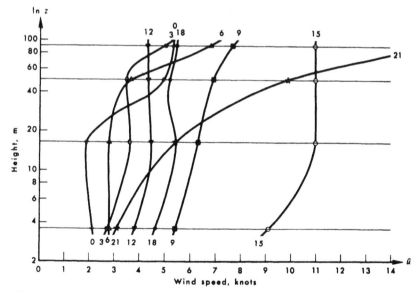

Fig. 6.6. Wind profiles between 3.6 and 90.9 m at Eglin Air Force Base, May 20, 1969. [After Hsu (1973).]

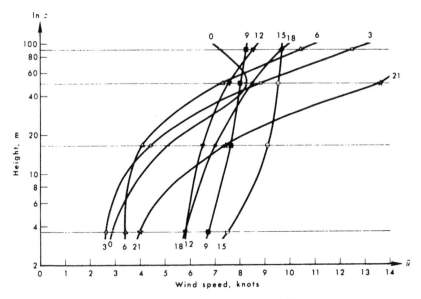

Fig. 6.7. Same as Fig. 6.6 for May 21, 1969.

between land breeze (when the wind blows from land to sea) and sea breeze (when the wind blows from sea to land) occurs.

The sea-breeze wind profile has been derived (Hsu, 1973) such that

$$U_z = 3\left(\frac{U_*^5 \bar{\theta}}{\lambda^3 g |\theta_*|}\right)^{1/3} (Z_0^{-1/3} - Z^{-1/3}) \tag{6.20}$$

where $\lambda \simeq 1$ (see, e.g., Deardorff and Willis, 1967).

Figure 6.8 shows some of the measured sea-breeze wind profiles. Also included in this figure is an example of the predicted sea-breeze wind profile using Eq. (6.20) and those parameters listed in the figure (for more detail, see Hsu, 1973).

A general formula for the diabatic wind profile can be derived (see, e.g., Panofsky and Dutton, 1984) so that, from Eq. (6.4),

$$\frac{\kappa Z}{U_*}\left(\frac{\partial U}{\partial Z}\right) = \phi_m\left(\frac{Z}{L}\right)$$

where, for simplicity, U_* has replaced U_{*0}. If one adds and subtracts 1 on the right of the above equation and integrates from the ground, where $Z = Z_0$ and $U = 0$, to an arbitrary height Z, one gets

$$U_z = \frac{U_*}{\kappa}\left[\ln\frac{Z}{Z_0} - \psi_m\left(\frac{Z}{L}\right)\right] \tag{6.21}$$

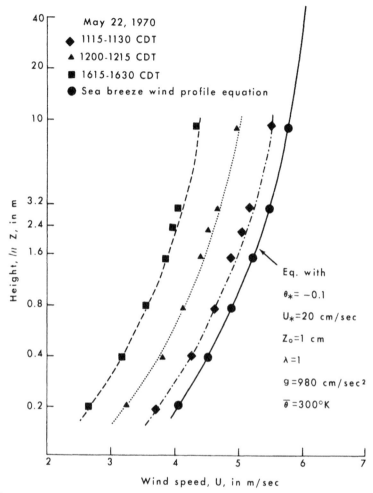

Fig. 6.8. Examples of measured sea-breeze wind profiles in the surface boundary layer near Fort Walton Beach. Included also is the predicted sea-breeze wind profile equation, with parameters from Eq. (6.20) specified in the figure (Hsu, 1973).

where

$$\psi_m\left(\frac{Z}{L}\right) = \int_{Z_0/L}^{Z/L} [1 - \phi_m(\zeta)] \frac{d\zeta}{\zeta}$$

where $\zeta = Z/L$.

In practice, Z_0/L is usually quite small, so that

$$\psi_m\left(\frac{Z}{L}\right) = \int_0^{Z/L} [1 - \phi_m(\zeta)] \frac{d\zeta}{\zeta} \tag{6.22}$$

Under unstable conditions Paulson (1970) provides that

$$\psi_m = \ln\left[\left(\frac{1 + X^2}{2}\right)\left(\frac{1 + X}{2}\right)^2\right] - 2\arctan X + \frac{\pi}{2} \qquad (6.23)$$

where $X = (1 - 16Z/L)^{1/4}$, and under stable conditions

$$\psi_m = -5\frac{Z}{L} \qquad (6.24)$$

6.4 Profiles of Temperature and Humidity

On the basis of the above discussion, similar expressions may be obtained for the temperature profile from Eq. (6.5), and for humidity from Eq. (6.6), such that (see, e.g., Panofsky and Dutton, 1984)

$$\frac{\theta - \theta_0}{T_*} = \frac{1}{\kappa}\left[\ln\frac{Z}{Z_*} - \psi_h\left(\frac{Z}{L}\right)\right] \qquad (6.25)$$

where, for unstable conditions,

$$\psi_h = 2\ln\{\tfrac{1}{2}[1 + (1 - 16Z/L)^{1/2}]\}$$

and for stable air,

$$\psi_h = -5\frac{Z}{L}$$

Note that when $-Z/L > 2$ these equations become inaccurate because the atmosphere is under the "free convection condition." The profile under sea breeze conditions as given in Eq. (6.20) is an example for wind, and similar derivation for temperature has been made (see Hsu, 1973).

Equations for humidity and other scalars may be derived similarly, such that

$$\frac{q - q_0}{q_*} = \frac{1}{\kappa}\left[\ln\frac{Z}{Z_q} - \psi_q\left(\frac{z}{L}\right)\right] \qquad (6.26)$$

Since very little information is available for ψ_q, at present it seems best to assume $\psi_q = \psi_h$ (Panofsky and Dutton, 1984).

Note that, similar to the determination of U_* as discussed in Section 6.2, under near-neutral conditions, when $Z/L = 0$, values of T_* and q_* can be obtained by the slope of the straight line fit, with $\ln Z$ as the axis and T or q as the ordinate. If so, values of τ, H, and E can be obtained from Eqs. (6.1)–(6.3).

If the condition is not near neutral, plots of $(\ln Z - \psi)$ versus θ or q must be made.

6.5   The Transition (or Ekman) Layer

Between the surface layer and the geostrophic wind level, a transition layer exists. If we assume, as did Hess (1959, p. 279), (1) horizontal mean motion, (2) horizontal mean wind shears are small compared to vertical mean wind shears, and (3) a balance between Coriolis, pressure gradient, and eddy viscosity forces at every level, the horizontal equations of motion similar to Eq. (4.32) may be written as

$$0 = -\frac{\partial P}{\partial X} + \rho f v + \frac{\partial \tau_{zx}}{\partial Z} \qquad (6.27)$$

$$0 = -\frac{\partial P}{\partial Y} - \rho f v + \frac{\partial \tau_{zy}}{\partial Z} \qquad (6.28)$$

Now, multiplying Eq. (6.28) by $i\,(=(-1)^{1/2})$ and adding Eq. (6.27), we have

$$0 = -\left(\frac{\partial P}{\partial X} + i\frac{\partial P}{\partial Y}\right) - \rho f i(u + iv) + \frac{\partial}{\partial Z}(\tau_{zx} + i\tau_{zy})$$

or

$$0 = -\left(\frac{\partial P}{\partial X} + i\frac{\partial P}{\partial Y}\right) - \rho f i(u + iv) + \rho k\frac{\partial^2}{\partial Z^2}(u + iv)$$

where ρk is an eddy exchange coefficient, which is independent of height.

If we orient the X axis parallel to the surface isobars with a positive geostrophic wind U_g, and further assume that pressure gradient and density do not vary appreciably with height so that

$$\partial P/\partial X = 0 \qquad \text{and} \qquad \partial P/\partial Y = -\rho f u_g$$

we get

$$\frac{d^2}{dZ^2}(u + iv - u_g) - \frac{if}{k}(u + iv - u_g) = 0 \qquad (6.29)$$

Equation (6.29) is a second-order, homogeneous, linear, differential equation with constant coefficient. The boundary conditions are

$$u + iv = 0 \qquad \text{at} \quad Z = 0 \qquad (6.30)$$

and

$$u + iv = u_g \qquad \text{at} \quad Z = \infty \qquad (6.31)$$

The solution to Eq. (6.29) may be written as

$$u + iv - u_g = Ae^{(if/k)^{1/2}z} + Be^{-(if/k)^{1/2}z} \qquad (6.32)$$

where A and B are arbitrary constants.

Since $(i)^{1/2} = (1 + i)/(2)^{1/2}$, Eq. (6.32) becomes

$$u + iv - u_g = Ae^{a(1+i)Z} + Be^{-a(1+i)Z} \tag{6.33}$$

where $a = (f/2k)^{1/2}$.

Now, from the boundary condition of Eq. (6.31), $A = 0$, and from Eq. (6.30), $B = -u_g$. Thus Eq. (6.33) becomes

$$u + iv - u_g = -u_g e^{-a(1+i)Z} = -u_g e^{-aZ} e^{-aiZ}$$

or

$$u + iv = u_g(1 - e^{-aZ} e^{-aiZ})$$

Since $e^{-iaZ} = \cos aZ - i \sin aZ$ (see, e.g., Hess, 1959, p. 228), we have

$$u + iv = u_g[1 - e^{-aZ}(\cos aZ - i \sin aZ)]$$
$$= u_g(1 - e^{-aZ} \cos aZ + i e^{-aZ} \sin aZ)$$
$$= u_g(1 - e^{-aZ} \cos aZ) + i u_g e^{-aZ} \sin aZ$$

Therefore, from real and imaginary parts,

$$u = u_g(1 - e^{-aZ} \cos aZ) \tag{6.34}$$

and

$$v = u_g e^{-aZ} \sin aZ \tag{6.35}$$

The solution to this set of equations [Eqs. (6.34) and (6.35)] can be seen graphically by means of a polar coordinate plot as a function of the height, such as shown in Figs. 6.9 and 6.10. This is called a hodograph, or Ekman spiral, after W. F. Ekman, who first obtained an analogous result for the

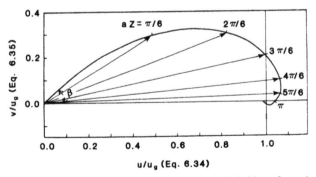

Fig. 6.9. The Ekman spiral represented by Eqs. (6.34) and (6.35). Note: β over land $> \beta$ over sea (see Fig. 4.4).

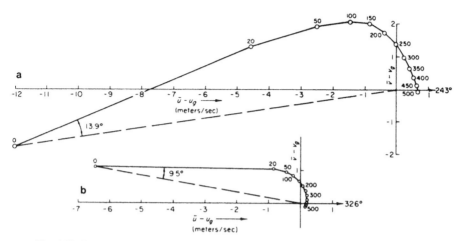

Fig. 6.10. Examples of vector differences of observed minus geostrophic wind as a function of height in the maritime friction layer near Scilly Isles. [Derived by Lettau (1957), from pilot balloon ascents of Sheppard *et al.* (1952).] Note that the top diagram (a) is for warm air over cold water where air is 0.46°C warmer than the water, whereas diagram (b) is for cold air over warm water, with air–sea temperature difference of −1.82°C. [After Roll (1965).]

surface layers of the ocean. Note that at $Z = \pi/a = \pi(2k/f)^{1/2}$ the v component will vanish for the first time and the wind will become parallel to the isobars. In the northern hemisphere the wind vector turns clockwise with elevation. Note also that the wind direction at or close to the surface makes a larger angle on land than offshore, with isobars and points toward the low pressure because there is more friction on land than on sea (cf. Fig. 4.4).

6.6   Surface Fluxes of Momentum, Heat, and Moisture

Formulas for computing the fluxes of momentum, heat, and moisture in the atmospheric surface layer have been provided in Eqs. (6.1), (6.2), and (6.3), respectively. They are obtained by the eddy correlation method, which is the most accurate and direct approach to estimating these fluxes. However, they are also difficult logistically for marine use. There is another approach that is much easier to apply in the real world but less accurate than the eddy correlation method. This practical method is called "bulk formula." The formulas are derived as follows:

From Eq. (6.21) we have

$$U_Z = (u_*/\kappa)[\ln(Z/Z_0) - \psi(\zeta)]$$

The difference between level 1 (lower elevation) and level 2 (upper elevation) is

$$u_2 - u_1 = \frac{u_*}{\kappa}[\ln(Z_2/Z_1) + \psi(\zeta_1) - \psi(\zeta_2)]$$

$$u_* = \frac{\kappa(U_2 - U_1)}{\ln(Z_2/Z_1) + \psi_M(Z/L)} \tag{6.36}$$

where $\psi_M(Z/L) = \psi(\zeta_1) - \psi(\zeta_2)$ and where u_* is the shear (friction) velocity.
    Similarly, from Eq. (6.25),

$$\theta_* = \frac{\kappa(\theta_2 - \theta_1)}{\ln(Z_2/Z_1) + \psi_H(Z/L)} \tag{6.37}$$

and, from Eq. (6.26),

$$q_* = \frac{\kappa(Q_2 - Q_1)}{\ln(Z_2/Z_1) + \psi_E(Z/L)} \tag{6.38}$$

Thus,

$$\tau/\rho = u_*^2 \qquad H/\rho C_p = -u_*\theta_* \qquad E/\rho = -u_*q_* \tag{6.39}$$

Setting $Z_1 = Z_0$ and $Z_2 = 10$ m, then

$$u_*^2 = \frac{\kappa^2 U_{10}^2}{[\ln(10/Z_0) + \psi_M(10/L)]^2} = C_D U_{10}^2 \tag{6.40}$$

or

$$\tau = \rho C_D U_{10}^2 \qquad \text{and} \qquad C_D = (u_*/U_{10})^2 \tag{6.41}$$

with $C_D = \kappa^2/[\ln(10/Z_0) + \psi_M(10/L)]^2$ the drag coefficient; the heat flux is
then

$$H/\rho C_p = -u_*\theta_*$$

$$= \frac{\kappa^2 U_{10}(\theta_0 - \theta_{10})}{[\ln(10/Z_0) + \psi_M(10/L)][\ln(10/Z_T) + \psi_H(10/L)]}$$

$$= C_H U_{10}(\theta_0 - \theta_{10}) \tag{6.42}$$

and, similarly, for moisture flux,

$$E/\rho = -u_*q_* = C_E U_{10}(Q_0 - Q_{10}) \tag{6.43}$$

Equations (6.41), (6.42), and (6.43) are the bulk methods for estimating the
momentum, heat, and moisture fluxes at the air–land or air–sea interfaces,
where C_D, C_H, and C_E are the transfer coefficients for the corresponding fluxes.
Values of these coefficients as measured from 1970 to 1975 and from 1973 to

1982 have been reviewed by Hsu (1978) and Blanc (1985), respectively. The most recent results are due to Large and Pond (1981 for momentum flux and 1982 for heat and moisture fluxes). They are summarized as follows.

For the drag coefficient,

$$C_D = C_{DN}\{1 + (C_{DN})^{1/2}\kappa^{-1}[\ln(Z/10) - \psi_M(Z/L)\}^{-2} \tag{6.44}$$

where C_{DN} is the drag coefficient under neutral stability conditions:

$$10^3 C_{DN} = \begin{cases} 1.14 & 4 \leq U_{10} < 11 \quad \text{m s}^{-1} \\ 0.49 + 0.065 U_{10}, & 11 \leq U_{10} \lesssim 25 \quad \text{m s}^{-1} \end{cases} \tag{6.45}$$

For heat and moisture transfer coefficients,

$$C_T = \frac{C_{TN}(C_D/C_{DN})^{1/2}}{1 + C_{TN}\kappa^{-1}C_{DN}^{-1/2}[\ln(Z/10) - \psi_h(Z/L)]}$$

$$C_E = \frac{C_{EN}(C_D/C_{DN})^{1/2}}{1 + C_{EN}\kappa^{-1}C_{DN}^{-1/2}[\ln(Z/10) - \psi_q(Z/L)]}$$

where

$$10^3 C_{TN} = \begin{cases} 1.13 & \text{for } Z/L < 0, \quad \text{unstable, } 4 < U_{10} < 25 \quad \text{m s}^{-1} \\ 0.66 & \text{for } Z/L > 0, \quad \text{stable, } 6 < U_{10} < 20 \quad \text{m s}^{-1} \end{cases} \tag{6.46}$$

$$10^3 C_{EN} = 1.15 \quad \text{for } Z/L < 0, \quad \text{unstable, } 4 < U_{10} < 14 \quad \text{m s}^{-1} \tag{6.47}$$

Functions of $\psi_m(Z/L)$, $\psi_h(Z/L)$, and $\psi_q(Z/L)$ are given in Sections 6.3 and 6.4, where (see Large and Pond, 1982)

$$Z/L \simeq \frac{-100Z}{U_z^2 T_0}(\Delta\theta + 1.7 \times 10^{-6}T_0^2 \Delta Q) \tag{6.48}$$

for $\Delta\theta > 0$, unstable, and

$$Z/L \simeq \frac{-70Z}{U_z^2 T_0}(\Delta\theta + 2.5 \times 10^{-6}T_0^2 \Delta Q) \tag{6.49}$$

for $\Delta\theta < 0$, stable, where

$$T_0 \simeq T_z(1 + 1.7 \times 10^{-6}T_z Q_z) \tag{6.50}$$

and, assuming a 75% relative humidity,

$$\Delta Q \simeq 0.98 Q_{Sat(T_s)} - 0.75 Q_{Sat(T_z)}$$

where $\Delta\theta = T_s - \theta_z$ and $\Delta Q = Q_s - Q_z$.

Note that the saturation humidity (g m^{-3}) over pure water at T degrees Kelvin is $Q_{Sat(T)} \simeq 64038 \times 10^4 \exp(-5107.4/T)$, and over salt water is $Q_s \approx 0.98 Q_{Sat(T_s)}$.

6.7 Wind Stress (Drag) Coefficient over Water Surfaces

Wind stress is one of the most important parameters for air–sea interaction studies [see, e.g., Roll (1965) for momentum flux; Bishop (1984) for upwelling (current) computation]. Many investigators have demonstrated that the wind stress or aerodynamic drag coefficient C_D increases with wind speed over water surfaces [see, e.g., Eq. (6.45)], whether they are small lakes or open oceans. A summary of these findings during the last 10 years is given in Fig. 6.11. Although it can be seen that C_D increases generally with wind speed, the rates of these increases vary greatly. Several efforts have been made to explain these variations, such as by Charnock's (1955) formulation (Garratt, 1977), by Froude number scaling (Wu, 1982), and by state of wave development (Donelan, 1982). A mechanism that can explain and estimate these variabilities satisfactorily (Hsu, 1986a) is presented in this section.

The wind stress or aerodynamic drag coefficient C_D is defined as [see Eq. (6.21)]

$$C_D = C_z = \left(\frac{U_*}{U_Z}\right)^2 = \left[\frac{\kappa}{\ln(Z/Z_0) - \psi_m(Z/L)}\right]^2 \qquad (6.51)$$

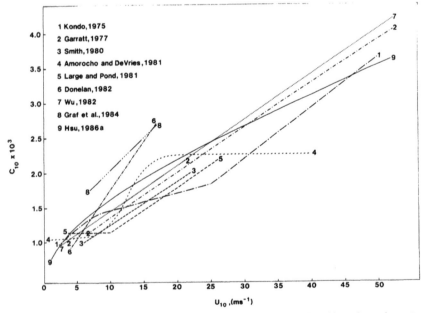

Fig. 6.11. Variation of the drag coefficient (C_{10}) with wind speed at 10 m above the water surface.

and under near-neutral conditions [see Eq. (6.15)]

$$C_{DN} = \left[\frac{\kappa}{\ln(Z/Z_0)}\right]^2 \tag{6.52}$$

Thus

$$C_D = C_{DN}\left[1 - \frac{\psi_m(Z/L)}{\ln(Z/Z_0)}\right]^{-2} \tag{6.53}$$

or

$$C_D = C_{DN}[1 - \kappa^{-1}C_{DN}^{1/2}\psi_m(Z/L)]^{-2} \tag{6.54}$$

Although the parameter Z_0 is obtained from integration of Eq. (6.4) when $U = 0$ at $Z = Z_0$ [see Eq. (6.21)], this parameter also has physical meaning. On the basis of K theory [e.g., Panofsky and Dutton, 1984; also cf. Eq. (4.42)],

$$\tau = \rho U_*^2 = \rho K_m \frac{\partial U}{\partial Z} \tag{6.55}$$

where K_m is the exchange coefficient of momentum, which is also called eddy viscosity.

Under steady-state and near-neutral conditions, Eq. (6.4) states that

$$\partial U/\partial Z = U_*/\kappa Z \tag{6.56}$$

According to Businger (1973), this expression indicates an infinite shear at the surface ($Z = 0$), which of course is unrealistic. In order to keep the shear finite, a surface roughness Z_0 is introduced such that

$$\partial U/\partial Z = U_*/\kappa(Z + Z_0) \tag{6.57}$$

Integration yields

$$U_z = \frac{U_*}{\kappa}\ln\left(\frac{Z + Z_0}{Z_0}\right) \tag{6.58}$$

Note that Eqs. (6.58) and (6.14) are the same. From Eqs. (6.55) and (6.57) we have

$$K_m = \kappa U_*(Z + Z_0)$$

and at the surface

$$K_{m,\,surface} = \kappa U_* Z_0 \tag{6.59}$$

According to Panofsky and Dutton (1984), Eq. (6.59) states that, at the surface, eddy viscosity is taken to represent the product of eddy size and eddy velocity, and we see now that Z_0 represents eddy size at the surface. Clearly, the rougher the ground, the larger these eddies can be. Thus Z_0 is a measure of surface roughness. It is thus called the roughness length. Measurements show that Z_0 varies from about 0.01 cm over ice or water to several meters over cities or irregular woods, illustrating that Z_0 is a measure of how efficiently momentum can be transferred into the ground at a given wind speed. Furthermore, since Z_0 represents eddy size at the surface, it depends not only on the height but also on shape and spacing of surface features.

Parameterization of Z_0 over the water surface was first formulated on dimensional considerations by Charnock (1955):

$$Z_0 = a(U_*^2/g) \tag{6.60}$$

where a is the Charnock coefficient, assumed to be a constant.

On the other hand, it is reasonable (see, e.g., Kitaigorodskii and Volkov, 1965; DeLeonibus and Simpson, 1972; Hsu, 1974a; Coantic, 1978; Donelan, 1982) to assume that Z_0 depends on the wave age C/U_* as well as upon the wave steepness H/L_w or

$$Z_0 = f\left(\frac{C}{U_*}, \frac{H}{L_w}\right) \tag{6.61}$$

where C is the wave celerity, H is the wave height, and L_w is the wave length. A relationship that exists between C/U_* and H/L_w (e.g., Brutsaert, 1982) states that for large wave age the wave steepness is small. For waves coming from deep water into shallow water, the wave age decreases and the wave steepness increases (e.g., Graf et al., 1984). Since the effect of atmospheric stability can be explicitly incorporated into the growth rate equation for surface gravity waves (Janssen and Komen, 1985), the stability parameter is not included in the parameterization of Z_0.

On the basis of these considerations and many experimental results, Hsu (1974a) proposed a relationship among these variables that includes both wind and wave contributions, such that

$$Z_0 = \frac{1}{2\pi}\left[\frac{H_{1/3}}{(C/U_*)^2}\right] \tag{6.62}$$

where $H_{1/3}$ is the significant wave height, defined as the average of the highest one-third waves, and $H_{1/3}$ and C depend implicitly on L_w:

$$H_{1/3} = 4\sigma \tag{6.63}$$

where

$$\sigma^2 = \int_0^\infty E(\omega)\,d\omega \tag{6.64}$$

in which $E(\omega)$ is the wave energy spectrum. Note that σ is the standard deviation of the wave record (see, e.g., U.S. Army Corps of Engineers, 1984). Further validation of Eq. (6.62) is provided by Graf *et al.* (1984).

Relationships of wave parameters before shoaling are

$$C = L_w/T \tag{6.65}$$

and

$$L_w = (g/2\pi)T^2 \tag{6.66}$$

so

$$C = (g/2\pi)T \tag{6.67}$$

where $T = f_m^{-1}$ is the wave period and f_m is the dominant frequency of the spectral peak.

Since waves are dependent on the duration t of the wind and its over-water trajectory, that is, the fetch (F), relationships among these parameters are needed. In practice, a simple method for making wave estimates is desirable, but is possible only if the geometry of the water body is relatively simple and the wave conditions are either fetch-limited or duration-limited. Under fetch-limited conditions, winds have blown constantly long enough for wave heights at the end of the fetch to reach equilibrium. Under duration-limited conditions, the wave heights are limited by the length of time the wind has blown. These two conditions represent asymptotic approximations to the general problem of wave growth. In most cases the wave growth pattern at a site is a combination of the two cases. Equations (6.68)–(6.73) are obtained from U.S. Army Corps of Engineers (1984) by simplifying the equation used to develop the parametric model based on JONSWAP experiments and others (Hasselmann *et al.*, 1976).

In the fetch-limited case, the parameters required are the fetch F and the adjusted wind speed U_A, as described in Chapter 3, Section IV, of U.S. Army Corps of Engineers (1984), and represent a relatively constant average value over the fetch. The spectral wave height H_{mo} and peak spectral period T_m are the parameters predicted.

$$\frac{gH_{mo}}{U_A^2} = 1.6 \times 10^{-3}\left(\frac{gF}{U_A^2}\right)^{1/2} \tag{6.68}$$

$$\frac{gT_m}{U_A} = 2.857 \times 10^{-1}\left(\frac{gF}{U_A^2}\right)^{1/3} \tag{6.69}$$

$$\frac{gt}{U_A} = 6.88 \times 10^1\left(\frac{gF}{U_A^2}\right)^{2/3} \tag{6.70}$$

Note that $T_{1/3}$ is given as $0.95T_m$. The preceding equations are valid up to the fully developed wave conditions given by

$$\frac{gH_{mo}}{U_A^2} = 2.433 \times 10^{-1} \tag{6.71}$$

$$\frac{gT_m}{U_A} = 8.134 \tag{6.72}$$

$$\frac{gt}{U_A} = 7.15 \times 10^4 \tag{6.73}$$

Note that a given calculation for a duration should be checked to ensure that it has not exceeded the maximum wave height or period possible for the given adjusted wind speed and fetch.

Examples of application of JONSWAP formulations are given in Graf et al. (1984), and in addition correction of wind difference between land and sea as well as temperature difference between sea and air are provided in the Liu et al. (1984). For our discussion we set $U_{10} = U_A$, $H_{1/3} = H_{mo}$, and $T = T_m$.

As a further check, since

$$T_{1/3} = U_{10}/0.13 \text{ g} \tag{6.74}$$

for fully developed sea, according to Pierson and Moskowitz (1964) (see Komen et al., 1984, Eq. 3.3),

$$T_m = T_{1/3}/0.95 \qquad \text{or} \qquad = U_{10}/0.95 \times 0.13 \text{ g} = 8.10 U_{10}/g$$

which is nearly equal to $8.13U_{10}/g$, as shown in Eq. (6.72)

From the Charnock equation as shown in Eq. (6.69) we see that Z_0 is related to the value of U_*, which by definition of Eq. (6.51) is related explicitly to the drag coefficient, wind speed, and atmospheric stability. However, the generalized roughness equation as proposed by Hsu (1974a) as shown in Eq. (6.62) explicitly incorporates the wave parameters in addition to the wind and stability because $H_{1/3}$ and C are related to the fetch, the duration, and the wind speed through Eqs. (6.67)–(6.73).

From Eqs. (6.60), (6.62), and (6.67) it can be shown that the Charnock coefficient a is

$$a = \frac{2\pi}{g} \left(\frac{H_{1/3}}{T^2} \right) \tag{6.75}$$

This equation shows that the Charnock coefficient is constant only when the ratio of $H_{1/3}/T^2$ is constant. For example, for fully developed seas, from Eqs. (6.71) and (6.72) we have

$$a = 0.023 \tag{6.76}$$

which is reasonable compared to the most recent value, 0.0185, obtained by Wu (1982). Note, however, that values of a reported in the literature vary widely, ranging from 0.0144 (Garratt, 1977) to 0.046 (Schwab, 1978). Variations in a are more likely due to field measurements made under conditions of partially developed seas. Furthermore, under fetch- or duration-limited conditions Eqs. (6.68) and (6.69) should be applied with Eq. (6.70) as a constraint.

On the basis of Eqs. (6.51), (6.62), and (6.67)–(6.73), a mechanism for the increase of drag coefficient with wind speed is proposed. The mechanism states that C_D is related explicitly to wind speed U_z, wave height $H_{1/3}$, phase velocity C, or wave period T, and atmospheric stability $\xi = Z/L$, such that

$$C_D = \kappa^2/[\ln Z - \ln Z_0 - \psi_m(\xi)]^2 \tag{6.51}$$

where Z_0 is given in Eq. (6.62).

Note that for a given $H_{1/3}$ and stability condition, the older the wave: that is, the larger C/U_*, the smaller the C_D value. This is explained by the following equation, derived by substituting Eq. (6.62) into Eq. (6.51):

$$C_D = \kappa^2/[\ln Z + \ln(2\pi) - \ln H_{1/3}$$
$$+ 2\ln(C/U_*) - \psi_m(\xi)]^2 \tag{6.77a}$$

This reasoning may explain the difference in the variation of C_D between lake environments such as Lake Ontario (Donelan, 1982) and Lake Geneva (Graf *et al.*, 1984) and oceanic environments (Smith, 1980; Large and Pond, 1981) (Fig. 6.11).

Since wave age is rarely measured, Z_0 may be reduced by incorporating Eqs. (6.67), (6.68), and (6.69) so that

$$Z_0 = AC_{10}F^{-1/6}U_{10}^{7/3} \tag{6.78}$$

where $A = 0.00859$.

Substituting Eq. (6.78) into Eq. (6.51) and simplifying;

$$C_D = \kappa^2/[\ln Z + 4.7571 - \ln C_{10} + 0.167\ln F$$
$$- 2.333\ln U_z - \psi_m(\xi)]^2 \tag{6.79}$$

Note that C_D depends more on the effect of wind speed than on fetch.

It is now clear that for a given wind speed and stability, when the fetch is large, C_D is small, as discussed above. On the other hand, for a given fetch and stability, the value of C_D will increase as the wind speed increases. Equation (6.79) is thus the mechanism proposed here to explain the increase of C_D with U_z. Note that its derivation is based on JONSWAP formulation [Eqs. (6.68)–(6.70)]. Therefore, to apply Eq. (6.79), Eqs. (6.68)–(6.70) should be validated. A quick way to do this is to use the nomogram of deepwater significant wave prediction curves as functions of wind speed, fetch length, and wind duration, as provided in U.S. Army Corps of Engineers (1984, Fig. 3-23 or 3-24).

On the basis of Eq. (6.77a), if the atmospheric stability term is neglected, since waves incorporate explicitly the effect of Z/L (see, e.g., Janssen and Komen, 1985), we have, by setting $Z = 10$ m,

$$C_{10} = \kappa^2/[\ln 10 + \ln(2\pi) - \ln H_{1/3}$$
$$+ 2\ln(C/U_*)]^2 \tag{6.77b}$$

One can also substitute and rearrange Eqs. (6.51), (6.67), (6.71), and (6.72) into Eq. (6.77b), thus bypassing by setting c/U_* as a constant, so that

$$0.4(C_{10}^{-1/2}) = 8.3538 - \ln C_{10} - 2\ln U_{10} \tag{6.77c}$$

Equation (6.77c) shows the relationship between U_{10} and C_{10} under fully developed sea conditions.

On the other hand, setting $C/U_* = 29$ for fully developed seas (see, e.g., Graf $et\ al.$, 1984) and substituting Eq. (6.71) into Eq. (6.77b), Eq. (6.77b) can be reduced directly to

$$C_{10} = \left(\frac{\kappa}{14.56 - 2\ln U_{10}}\right)^2 = \left(\frac{0.4}{14.56 - 2\ln U_{10}}\right)^2 \tag{6.80}$$

Both Eqs. (6.77c) and (6.80) are plotted in Fig. 6.12. It can be seen that the difference is not great for practical applications. Because Eq. (6.77c) is more difficult to use than Eq. (6.80), Eq. (6.80) is recommended for operational use.

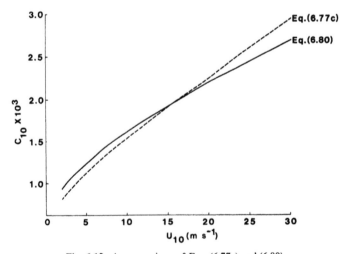

Fig. 6.12. A comparison of Eqs. (6.77c) and (6.80).

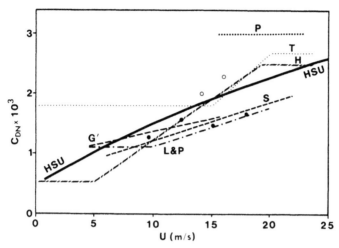

Fig. 6.13. The neutral drag coefficient C_{DN} versus wind speed U from other sources. The solid lines are regression lines from eddy correlation estimates of $\overline{u'w'}$; the dashed lines are the formulae adopted by three storm-surge modelers; the solid circles are derived from the water-level fluctuations over several months (Schwab, 1981); the open circles are derived from the peak storm surge for two storms (Simons, 1974, 1975). The curve labeled HSU is based on Eq. (6.80). See text for explanation. [After Donelan (1982).]

In order to compare Eq. (6.80) with other studies, it is plotted as line number 9 in Fig. 6.11. It can be seen that C_{10} increases with wind speed and that the value of C_{10} calculated by the present method is approximately the mean value found by the other methods. As a cross check, under the condition of fully developed seas, that is, in a region of nearly unlimited fetch and duration such as the tropics, wind speed usually ranges from 5 to 10 m s^{-1}. Equation (6.80) predicts that C_{10} ranges from 1.2 to 1.6×10^{-3}, which is in good agreement with the constant value of 1.5×10^{-3} from BOMEX (Pond *et al.*, 1971) and $1.4 (\pm 0.4) \times 10^{-3}$ from GATE (Businger and Seguin, 1977).

As a further test of Eq. (6.80) against other sources, Fig. 6.13 is provided. Except for the curve labeled HSU, the figure is the same as in Donelan (1982, Fig. 2). In the figure G' is obtained from Garratt (1977), S from Smith (1980), L & P from Large and Pond (1981), H from Heaps (1969), T from Timmerman (1977), and P from Platzman (1963). The lines indicated by G', S, and L & P are regression lines from eddy correlation estimates; the lines representing T, H, and P are the formulas adopted by three storm-surge modelers; the solid circles are derived from water-level fluctuation over several months (Schwab, 1981); and the open circles are derived from the peak storm surges for two storms (Simons, 1974, 1975). From Fig. 6.13 it is concluded that our Eq. (6.80) may be considered as an average line, which is certainly consistent with the sources discussed above. Therefore, Eq. (6.80) should be a useful formula for practical applications in the marine environment.

When the sea is not fully developed, both fetch and duration of the wind should be taken into account and Eqs. (6.68)–(6.73) should be employed. Because Eq. (6.79) is transcendental, an iteration scheme may be applied. The term C_{10} converges after 5–10 iterations, typically. For practical applications, a nomograph to determine U_* from wind and wave parameters has been provided in Hsu (1976), as shown in Fig. 6.14. The equation in the nomograph is obtained by substituting Z_0 from Eq. (6.62) into Eq. (6.51). Note that since waves incorporate explicitly the effect of atmospheric stability (Janssen and Komen, 1985), Z/L is not involved in this nomograph. Figure 6.14 has been applied successfully under both adiabatic and nonadiabatic conditions for estimating fluxes of momentum (Hsu, 1976) and heat (Hsu, 1983c) over the ocean.

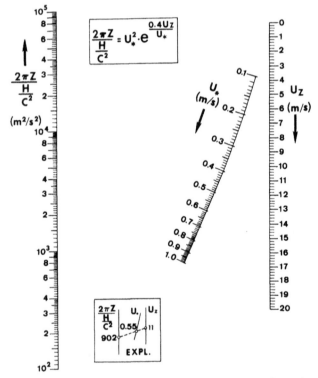

Fig. 6.14. Determining U_* from both wind and wave parameters. The figure gives an example of the use of commonly available wind (U_z) and wave (H and C) parameters to obtain U_*. [After Hsu (1976). Copyright by D. Reidel Publishing Co. Reprinted by permission.]

6.8   Some Aspects of Air–Sea Interactions across Ocean Fronts

The sea-surface temperature (SST) offshore from the coast is seldom homogeneous, particularly when fronts such as the Gulf Stream or the Kuroshio meander along coastal regions of, for example, Florida, North Carolina, and Korea. Some of their effects on the atmospheric boundary layer, and vice versa, that is, air–sea interaction, are presented in this section.

6.8.1   Internal Boundary Layer over Warm Currents

Significant atmospheric and oceanographic differences have often been observed in the region lying on either side of an oceanic front with a sharp temperature gradient such as the Gulf Stream (Sweet *et al.*, 1981; LaViolette, 1982) or the Kuroshio Current (Mahrt and Paumier, 1982). Of the many geophysical aspects associated with these differences, one is the development of a so-called internal boundary layer (IBL) over the warm-water side when the wind blows at right angles from the cold-water side (Fig. 6.15).

The oceanic thermal front may be considered a transitional zone in which the air flow adjusts to a new set of boundary conditions when it crosses the front. In the process, air adjacent to the sea surface becomes modified. This modified layer is often called the internal boundary layer (Fig. 6.15).

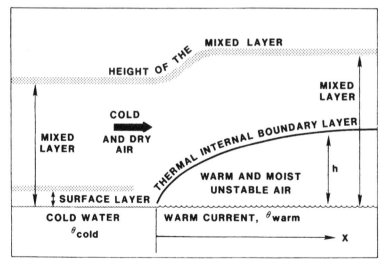

Fig. 6.15. Schematic diagram of the thermally modified boundary layer, or the internal boundary layer, over a warm oceanic current caused by advection of cold air (for explanation of symbols, see text).

Fig. 6.16. Warm oceanic current (dark area) in the Korea Strait. Ship track is outlined. NOAA-6 meteorological satellite image (November 17, 1980). Channel 4, calibrated to 0.25°C resolution, was used. (Courtesy of Oscar K. Huh of the Coastal Studies Institute, Louisiana State University.)

Venkatram (1977) has derived a theoretical formula from the first principles that states that

$$h = \frac{U_*}{U_m}\left[\frac{2(\theta_{warm} - \theta_{cold})X}{\gamma(1 - 2F)}\right]^{1/2} \tag{6.81}$$

where h is the height of the IBL, U_* and U_m are the friction velocity and mean wind speeds, respectively, inside the IBL, γ is the lapse rate above the boundary layer or upwind condition, F is an entrainment coefficient, which ranges from 0 to 0.22, θ_{warm} and θ_{cold} are the potential temperatures over the warm- and cold-water sides, respectively, and X is the distance or fetch downwind from the front.

Equation (6.81) may be simplified by employing the drag coefficient, which is defined as $(U_*/U_m)^2$ [see Eq. (6.41)]. Thus

$$h = \left[\frac{2C_D(\theta_{warm} - \theta_{cold})X}{\gamma(1 - 2F)}\right]^{1/2} \qquad (6.82)$$

Equation (6.82) will be shown to be verified by observations.

To provide an example, the Korea Strait is chosen. Because of the monsoonal effect, the prevailing wind direction during fall and winter over the Korea Strait is northwesterly. The speed varies from 5 to 15 m s^{-1}. The winds are approximately perpendicular to the oceanic thermal front associated with the Tsushima current, a branch of the Kuroshio.

The warm oceanic current in the Korea Strait is readily seen in satellite imagery (Fig. 6.16). Dry-bulb (T_{dry}) and dew-point (T_{dew}) temperatures over the cold-water side are shown in the left panel in Fig. 6.17, and those over the warm side are given on the right. The measurements were made by radiosondings from ship track line E (Fig. 6.16). There is an approximately 30% increase in relative humidity and mixing ratio from the cold side to the

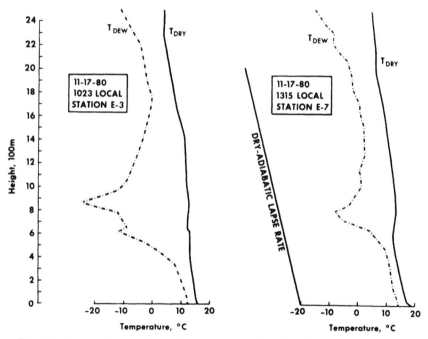

Fig. 6.17. High-resolution radiosondings over the cold (left) and warm (right) sides of the oceanic current in the Korea Strait (cf. Fig. 6.16). Note that station E7 is located (in Fig. 6.16) near the lower left corner of letter E on line E, which was 15 km downwind from the oceanic front. The location of this station was 33°38′N and 128°00′E. [After Hsu (1984a).]

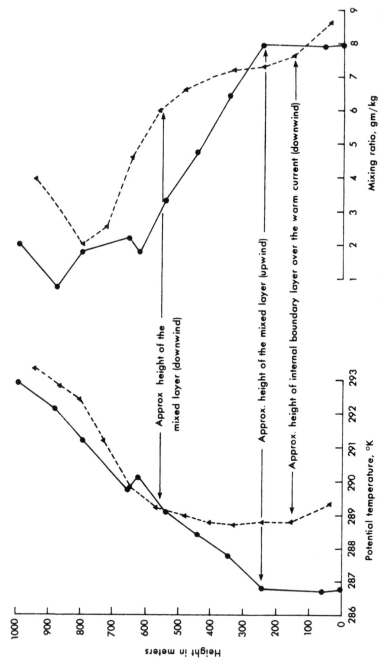

Fig. 6.18. Radiosondes from the Korea Strait. Profiles of potential temperature and mixing ratio for November 17, 1980, on both upwind and downwind sides of the thermal front (cf. Fig. 6.16) (see text for explanation). ●, station E-3, over cold water upwind, 1023 local time. ▲, station E-7, over warm current downwind, 1315 local time.

warmer side about 15 km downwind from the ocean front. As expected, the modification occurred in the lower atmosphere.

The height of the mixed layer is lower over the cold side than over the warm side, where the effect of surface heat flux is larger (Fig. 6.18). The mixing height increases from approximately 250 m on the upwind (cold-water) side to more than 550 m on the warm-water side. The corresponding water temperature is 11.5°C on the cold side and 18.0°C on the warm side. Note that the determination of the mixing height is based on both profiles of the potential temperature and the mixing ratio. Since the data points or sampling rates are not dense enough, these determinations are not absolute. More detailed high-resolution radiosondings were obtained in a followup experiment. An example is shown in Fig. 6.19. Comparison of Fig. 6.19 with results obtained by Wyngaard et al. (1978) indicates that, except for the atmospheric structure at around 250 m, this figure is about the same as theirs.

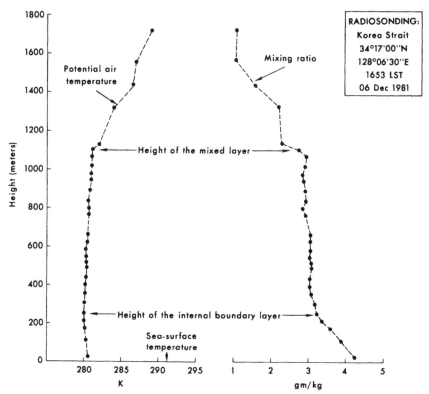

Fig. 6.19. Profiles of potential temperature and mixing ratio for December 6, 1981, over the warm side of the Tsushima Current in the Korea Strait. [After Hsu (1984a).]

As shown in Figs. 6.18 and 6.19, the precision of determination of the IBL height is dependent on the sampling rate. However, the accuracy of the IBL height (see, e.g., 150 m in Fig. 6.18) is related mainly to the pressure measurement near the surface. Since the pressure accuracy in the surface layer is less than 1 mb, the variation is therefore less than ± 10 m at 150 m. This is based on application of the standard hypsometric equation.

The increase in mixing height over the warm-water side has also been observed by Sweet *et al.* (1981). They showed that a 320-m increase occurs when the air crosses an 8°C front and a 400-m increase occurs when it crosses a 12°C oceanic thermal front. The slope changes around 150 m, shown in Fig. 6.18, and 250 m, in Fig. 6.19, cannot be explained by the physics of the surface layer or constant flux layer, which is normally of the order of 10 m over the water (see, e.g., Danard, 1981). Note also that under superadiabatic or free convection conditions in a desert environment the convective surface boundary layer over land can reach only ~ 100 m, as discussed by Danard (1981) and measured by Hsu (1983b).

On the basis of observations discussed above, Eq. (6.82) may be applied. Since the Kuroshio is a warm oceanic current originating in the tropics, we adopt $C_D = 0.0015$ (Pond et al., 1971). Figure 6.18 shows that $(\theta_{\text{warm}} - \theta_{\text{cold}}) = 2.7$°C. A common value of F is 0.2 (e.g., Driedonks, 1982). From the ship's log and Fig. 6.18, we have $X = 15$ km or 15,000 m and $\gamma = 1$°C/100 m.

Substituting these values into Eq. (6.22), we obtain $h = 142$ m. This computed value is in excellent agreement with the measured value of 150 m, as shown in Fig. 6.18 (the accuracy of this measured value of 150 m has been discussed previously). Note that the height to fetch ratio, h/X, is $\sim 1/100$, which is consistent with values discussed in the literature (Oke, 1978).

6.8.2   Atmospheric Mixing Height across Thermal Fronts

When temperature discontinuity exists, such as in a region influenced by an ocean thermal front (e.g., Gulf Stream or Kuroshio), the atmospheric boundary layer (ABL) undergoes certain changes. These adjustments include variations in mixing height (Figs. 6.15 and 6.20), the generation of sea-breeze-like winds (see Section 6.8.3) when the geostrophic wind is weak (say, <5 m s^{-1}), or the development of an internal boundary layer when synoptic-scale winds are stronger (≥ 5 m s^{-1}) (see Section 6.8.1).

Since the concentration of water vapor and aerosols within the marine atmospheric boundary layer is affected directly by the thickness of this boundary layer, it is important to study the variation of the mixing height in order to improve correction for vapor effects on SST in infrared images obtained remotely by airplane or satellite. Another application of this study is

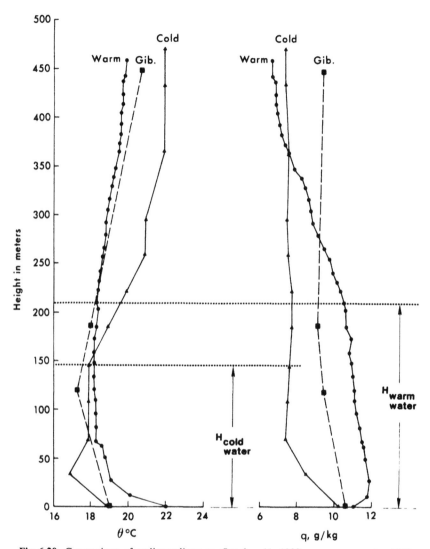

Fig. 6.20. Comparison of radiosondings on October 11, 1982, over warm water (SST = 20.4°C) at 1115Z and cold water (SST = 16.9°C) at 1013Z of the Alboran Sea as well as from Gibraltar (Gib.) at 1100Z. The symbol θ is the potential temperature and q is the mixing ratio. Note the difference in boundary layer height between cold and warm water.

to provide an operational forecasting model of the ducting height for radiowave propagation across the front if the mixing height on one side of the front is available (see Section 8.4).

In the atmospheric boundary layer, diffusion of some identifiable property such as potential temperature θ may be written as (see, e.g., Hess, 1959, p. 284)

$$\frac{\partial \bar{\theta}}{\partial t} = -\left(\bar{u}\frac{\partial \bar{\theta}}{\partial x} + \bar{v}\frac{\partial \bar{\theta}}{\partial y} + \bar{w}\frac{\partial \bar{\theta}}{\partial z}\right)$$

$$-\frac{1}{\rho}\left[\frac{\partial}{\partial x}(\rho\overline{\theta'u'}) + \frac{\partial}{\partial y}(\rho\overline{\theta'v'}) + \frac{\partial}{\partial z}(\rho\overline{\theta'w'})\right]$$

$$\pm Q \qquad \qquad (6.83)$$

where u, v, and w are the velocity components of the wind in the x, y, and z directions, respectively, ρ is the density, overbars and primes are the mean and fluctuating quantities, respectively, and Q is the rate of change following the air motion caused by processes other than turbulent transfer, such as radiative divergence.

Equation (6.83) states that the local rate of change of the mean value of θ is produced by three effects. First is the mean advection of θ, second is convergence of the eddy transport of θ by turbulent wind, and third is the source (production) or sink (destruction) of θ in question, if it occurs in the boundary layer considered.

In the atmospheric boundary layer, variation of ρ is small, and vertical transport is normally much larger than horizontal transport. Thus, Eq. (6.83) may be simplified as

$$\frac{d\bar{\theta}}{dt} \pm Q = -\frac{\partial}{\partial z}(\overline{\theta'w'}) \qquad \qquad (6.84)$$

where $d\bar{\theta}/dt = \partial\bar{\theta}/\partial t + \bar{u}(\partial\bar{\theta}/\partial x) + \bar{v}(\partial\bar{\theta}/\partial y) + \bar{w}(\partial\bar{\theta}/\partial z)$ (that is, the total derivative of $\bar{\theta}$ is made up of the sum of the local and convective derivatives of $\bar{\theta}$) (see, e.g., Hess, 1959).

Integrating Eq. (6.83) from the sea surface, where $Z = 0$, to $Z = H$, the top of the mixed layer, yields

$$H = \frac{(\overline{\theta'w'})_0 - (\overline{\theta'w'})_H}{\langle(d\theta/dt) \pm Q\rangle} \qquad \qquad (6.85)$$

where $\langle(d\theta/dt) \pm Q\rangle$ is an appropriate mean value in the mixed layer.

Equation (6.85) has been applied extensively in atmosphere boundary layer research (see, e.g., Tennekes and Driedonks, 1981). In the convective mixed layer

$$-(\overline{\theta'w'})_H = C(\overline{\theta'w'})_0 \qquad \qquad (6.86)$$

where C is the entrainment constant. The values reported for C range between 0 and 1, with an average value of 0.2 (Stull, 1976). Driedonks (1982) illustrated that a change in C from 0.2 to 0.5 leads to a variation of 20% in H, which may be considered within the accuracy of actual observations.

Substituting Eq. (6.86) into Eq. (6.85), we have

$$H = \frac{(1 + C)(\overline{\theta'w'})_0}{\langle (d\theta/dt) \pm Q \rangle} \tag{6.87}$$

The sensible heat flux $(\overline{\theta'w'})_0$ in the atmospheric boundary layer over the ocean may be written as (e.g., Pond et al., 1971)

$$(\overline{\theta'w'})_0 = C_T u(T_{sea} - T_{air}) \tag{6.88}$$

where C_T is the transfer coefficient for sensible heat, which is around 1.0×10^{-3} (see Large and Pond, 1982), and T_{sea} and T_{air} are temperatures for sea and air, respectively. Note that in the atmospheric layer, near the surface, pressure is around 1000 mb and thus θ may be replaced by T (e.g., Stage and Businger, 1981).

Substituting Eq. (6.88) into Eq. (6.87), one gets

$$H = \frac{(1 + C)C_T u(T_{sea} - T_{air})}{\langle (d\theta/dt) \pm Q \rangle} \tag{6.89}$$

Assuming that Eq. (6.89) is valid on both sides of the oceanic thermal front and that the sea-surface temperature (SST) across the front is the most important variable as compared to others in Eq. (6.89), we have, from a statistical point of view,

$$\Delta H \propto \Delta T \tag{6.90}$$

or

$$\Delta H = K \Delta T \tag{6.91}$$

or

$$\Delta H = a + b \Delta T \tag{6.92}$$

where ΔH (in meters) is the difference in the mixing height from warm water to cold water, ΔT (in °C) is the difference in SST across the front, and K (in m °C^{-1}) is a parameter related to the entrainment coefficient, wind speed, transfer coefficient for sensible heat flux, local change of the potential temperature in the mixed layer, advection due to winds, and radiative processes due to clouds (Hsu et al., 1985). Note that the coefficients a, b, and K can be obtained from statistical analysis.

Atmospheric soundings across oceanic thermal fronts were made from the Gulf Stream by Sweet et al. (1981) and from the Tsushima Current in the Korea Strait, a branch of the Kuroshio, by Hsu (1982a). According to Sweet et al. (1981), a 320-m increase occurred when the air crossed an 8°C front going from the cold to the warm water, and a 400-m increase occurred when air crossed a 12°C front. In the Korea Strait, the mixing height increased from 340 to 480 m when cold air crossed from the cold-water side of a 6°C front to the warmer side (Hsu, 1982a). However, three pairs of soundings are not enough for quantitative deduction; more pairs are needed. In this regard, additional studies were made as part of an October 1982 field experiment in the Alboran Sea, which is located between Spain and Morocco. The gyre that typifies the surface circulation in the Alboran Sea can be seen in the thermal analysis of satellite images and aircraft data, as shown in Figs. 2.4 and 2.5, respectively. During most of our experimental period, particularly on October 11, 12, and 16, 1983, the area was mainly under the influence of the Azores high-pressure system. Therefore, skies were mostly clear and winds were light during these days, as illustrated by the satellite images in Fig. 2.4. An example of light winds at 300 m above the Alboran Sea is shown in Fig. 2.5.

Pairs of high-resolution radiosondes were launched from two ships. These radiosonding pairs were made within 2 hr of each other. An example is shown

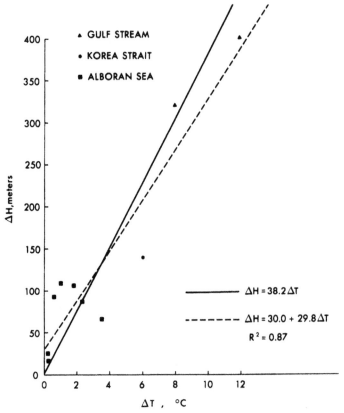

Fig. 6.21. Experimental results for ΔH versus ΔT as guided by Eq. (6.91) for the three geographic regions studied (see text for explanation). [From Hsu *et al.* (1985).]

in Fig. 6.20. Note that the mixing height is determined at the level where the first major change in the slope of either mixing ratio or potential temperature profile occurred. This technique has been applied commonly to the study of atmospheric boundary layers (e.g., Wyngaard *et al.*, 1978). Figure 6.20 also shows that the standard radiosonde measurement over land even from a small peninsula such as Gibraltar cannot be used to represent detailed marine boundary-layer structures.

Pairs of radiosondings obtained from the Gulf Stream, Korea Strait, and Alboran Sea are plotted in Fig. 6.21. Two statistical lines are drawn as guided by Eqs. (6.91) and (6.92). The linear regression for all pairs is [cf. Eq. (6.92)]

$$\Delta H = 30.0 + 29.8\,\Delta T \tag{6.93}$$

with the coefficient of determination $R^2 = 0.87$. In Eq. (6.93), ΔH is in meters and ΔT is in degrees Celsius. However, the average of all pairs indicates that [cf. Eq. (6.91)]

$$\Delta H = 38.2\,\Delta T \tag{6.94}$$

Equation (6.94) is determined by connecting the mean value and the origin. Based on the standard error of estimate, the 95% confidence level indicates that the intercept of Eq. (6.93) is not significantly different from zero. Therefore, Eq. (6.94) is applicable statistically. In addition, since the difference between these two lines is within experimental error (say, around 40 m for $\Delta H = 400$ m and $\Delta T = 10°C$), one may conclude that our theoretical equation, Eq. (6.91), is useful as a good first approximation for the oceanic thermal frontal environment.

6.8.3 Thermally Induced Winds across the Gulf Stream

According to Sweet *et al.* (1981), significant atmospheric and sea state differences have often been observed in regions lying on either side of the north wall of the Gulf Stream. A description of the Gulf Stream has been provided by Stommel (1966). Fog and haze, with poor visibility, frequently typify conditions on the cold shelf-water side, while convective phenomena (in some cases thunderstorms), with generally improved visibility, are often characteristic of areas on the warm side of the north wall. As shown in Fig. 6.22, winds at altitudes of 61–610 m over cold water varied from 10 to 20 knots from the southwest. A slight veering of the wind to the south-southwest occurred during the measuring aircraft's crossing of the north wall of the Gulf Stream, and the wind speed increased abruptly to 25–35 knots. Since the synoptic scale effect is small, this mesoscale phenomenon is very similar to the classical sea-breeze circulation (see, e.g., Hsu, 1970; Atkinson, 1981).

The motivation of this section is to provide a simple analytical model to explain these sea-breeze-like winds that blow from the cool-water side to the

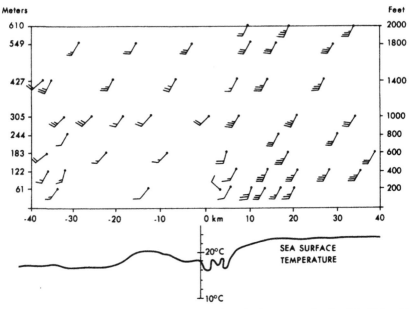

Fig. 6.22. Winds as determined from the inertial navigation system for the eight flight altitudes, December 6, 1979. One full bar equal $\sim 5 \text{ m s}^{-1}$ (10 knots). [After Sweet *et al.* (1981).]

warm-water side of the Gulf Stream. This section should be compared to that by Sweet *et al.* (1981), inasmuch as observational aspects were presented in that paper.

The sea-breeze-like circulation model to be considered in this study (see Fig. 6.23) takes friction into account (Hsu, 1984b). We assume, as did Hauwritz (1947), that the circulation takes place in a vertical x, z plane.

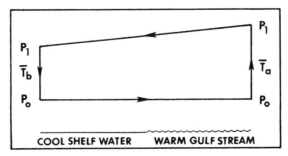

Fig. 6.23. Path of integration of the sea-breeze-like circulation across the Gulf Stream. [From Hsu (1984b).]

The x axis is chosen to be in the direction of the wind, which blows from the cool shelf water to the warm water across the oceanic thermal front in the Gulf Stream. The frictional force is assumed to be opposite to, and proportional to, the wind velocity. With these assumptions, the equations of motion for this sea-breeze-like wind system can be written in the following form:

$$\frac{du}{dt} + ku = -\frac{1}{\rho}\left(\frac{\partial p}{\partial x}\right) + fv \tag{6.95}$$

$$\frac{dv}{dt} + kv = -\frac{1}{\rho}\left(\frac{\partial p}{\partial y}\right) - fu \tag{6.96}$$

$$\frac{dw}{dt} + kw = -\frac{1}{\rho}\left(\frac{\partial p}{\partial z}\right) - g, \tag{6.97}$$

where u, v, and w are the velocity components in x, y, and z directions, respectively, p is the pressure, ρ is the density, g is the acceleration of gravity, and k is a constant that expresses the intensity of the frictional force.

From circulation theorem (e.g., Hess, 1959, p. 238), we have

$$c = \oint (u\,dx + v\,dy + w\,dz) \tag{6.98}$$

By multiplying Eq. (6.95) by dx, Eq. (6.96) by dy, Eq. (6.97) by dz, adding, and integrating around the closed path, one derives the rate of change of the circulation c to be

$$
\begin{aligned}
\frac{dc}{dt} &= \oint \left(\frac{du}{dt}dx + \frac{dv}{dt}dy + \frac{dw}{dt}dz\right) \\
&= \oint \left\{ \left[-ku + fv - \frac{1}{\rho}\left(\frac{\partial p}{\partial x}\right)\right]dx \right. \\
&\quad + \left[-kv - fu - \frac{1}{\rho}\left(\frac{\partial p}{\partial x}\right)\right]dy \\
&\quad \left. + \left[-kw - g - \frac{1}{\rho}\left(\frac{\partial p}{\partial z}\right)\right]dz \right\} \\
&= \oint (-ku\,dx - kv\,dy - kw\,dz) \\
&\quad + \oint (fv\,dx - fu\,dy) + \oint (-g)\,dz \\
&\quad + \oint \left(-\frac{1}{\rho}\right)\left(\frac{\partial p}{\partial x}dx + \frac{\partial p}{\partial y}dy + \frac{\partial p}{\partial z}dz\right)
\end{aligned}
$$

Hence, from Eq. (6.98),

$$dc/dt = -kc + f \oint (v\,dx - u\,dy) - \oint g\,dz - \oint dp/\rho \qquad (6.99)$$

The term $f \oint (v\,dx - u\,dy)$ represents the Coriolis effect, in which $u\,dy$ is a contribution to a rate of expansion of the enclosed area, while $v\,dx$ is a contribution to a rate of contraction of this area (Hess, 1959, p. 240). If one assumes that the rate of expansion and that of contraction balance each other, that is,

$$f \oint (v\,dx - u\,dy) = f \oint \left(\frac{dy}{dt}dx - \frac{dx}{dt}dy\right) = 0$$

The integral $\oint g\,dx$ is zero because it is the closed line integral of an exact differential, assuming g to be a single-value function of z.

Using the ideal gas law (i.e., $p = \rho RT$, where R is the gas constant for air and T is the air temperature), the term

$$-\oint_{p_0}^{p_1} \frac{dp}{\rho} = -R\int_{p_0}^{p_1} T\,dp/p$$

$$= -R\bar{T}\int_{p_0}^{p_1} dp/p = R\bar{T}\ln p_0/p_1$$

where $\bar{T}$ is a layer-mean temperature.

In the sea-breeze problem, $\bar{T}$ over land ($\bar{T}_a$ in Fig. 6.23) may be represented by superadiabatic lapse rate (Hsu, 1973) and $\bar{T}$ over water ($\bar{T}_b$) by adiabatic lapse rate because the air–sea temperature difference is smaller over water than over land (see, e.g., Roll, 1965). By analogy, $\bar{T}_a$ over the warm side (cf. Fig. 6.23) may be thought of as $\bar{T}$ over warm land, and $\bar{T}_b$ over the cold side may be thought of as $\bar{T}$ over water. Therefore, as a first approximation, $\bar{T}_a$ may be represented by superadiabatic lapse rate, whereas $\bar{T}_b$ will be represented by adiabatic lapse rate. Since the air temperature at p_1 is assumed to be nearly the same on both sides, the air temperature near the surface or the sea-surface temperature may be substituted for the vertically averaged $\bar{T}_a$ and $\bar{T}_b$.

On the basis of this reasoning, Eq. (6.99) may be written in the form

$$dc/dt = R(\bar{T}_a - \bar{T}_b)\ln(p_0/p_1) - kc \qquad (6.100)$$

following Hauwritz (1947); Eq. (6.98) can also be approximated by

$$c = L\bar{V} \qquad (6.101)$$

where L is the length of the path of integration, as shown in Fig. 6.23, and p_0 is the pressure near the surface and p_1 is the maximum height due to vertical convection. We use the mixing height in the afternoon here as a first approximation. The term $\bar{V}$ is the mean speed of the sea-breeze-like circulation along the path of integration. Then, from Eqs. (6.100) and (6.101) one gets

$$\frac{d\bar{V}}{dt} + k\bar{V} = (\bar{T}_a - \bar{T}_b)\frac{R}{L}\ln\frac{p_0}{p_1} = M(\bar{T}_a - \bar{T}_b) \qquad (6.102)$$

Assuming $(\bar{T}_a - \bar{T}_b)$ varies on a time scale much longer than k^{-1}, it follows that $d\bar{V}/dt$ may be neglected in Eq. (6.102), so that

$$\bar{V} = (M/k)(\bar{T}_a - \bar{T}_b) \qquad (6.103)$$

Thus, in the sea-breeze-like area, if the temperature difference between the air over the Gulf Stream and that over the cool shelf water and the path of integration are known, Eq. (6.103) may be used to calculate the mean speed of the sea-breeze circulation.

In order to test the model, observational data collected by Sweet et al. (1981) on December 6, 1979, are employed (cf. Fig. 6.22), and the following choices for the model parameters are made:

(a) $p_0 = 1017$ mb. This is based on Fig. 8 in Sweet et al. (1981).

(b) $p_1 = 890$ mb. This is based on Fig. 12a in Sweet et al. (1981).

(c) $L = 4 \times 150 = 600$ km. It is assumed that the average width of the Gulf Stream is 150 km (Stommel, 1966). About one-fourth of the total path of integration in Fig. 6.23 lies on the warm side of the north wall [for the method of obtaining total path of integration, see Hsu (1970)]. Note that the total path of integration is the sum of lengths $\overline{p_0p_1} + \overline{p_1p_1} + \overline{p_1p_0} + \overline{p_0p_0}$ (Fig. 6.23). Because $\overline{p_0p_1}$ and $\overline{p_1p_0}$ are small, we approximate the total path at twice $\overline{p_0p_0}$ or four times half of $\overline{p_0p_0}$, that is, the width of the Gulf Stream on the warm side of the north wall.

(d) It is assumed that the entire width of the warm Gulf Stream acts as the heat source for the solenoids in this kind of circulation system. Here $\bar{T}_a - \bar{T}_b = 4°C$. The air temperature from cross sections on the warmer side (22°C) and that on the cool side (18°C) were used (Sweet et al., 1981, Fig. 12).

(e) $k = 3.2 \times 10^{-5}$ s^{-1}. This is based on $k = f\tan\psi$, where f is the Coriolis parameter ($f = 2\Omega\sin\phi$, where Ω is the angular velocity for the earth's rotation and ϕ is the latitude) and ψ is the angle between the wind near the surface and the geostrophic wind (see Wallace and Hobbs, 1977, p. 378). At latitude 37°N, $f = 8.77 \times 10^{-5}$ s^{-1}. The parameter ψ over the study area was obtained between 10 and 30 degrees (Sweet et al., 1981, Fig. 8) with an average of 20 degrees. Therefore $k = 8.77 \times 10^{-5}\tan\psi = 3.2 \times 10^{-5}$ s^{-1}.

With the above parametric values we get

$$M = 0.0064 \quad \text{cm s}^{-2}\text{ deg}^{-1}$$

and

$$\bar{V} = M(\bar{T}_a - \bar{T}_b)k^{-1}$$
$$= 2.0(\bar{T}_a - \bar{T}_b) \quad \deg^{-1} \text{m s}^{-1}. \tag{6.104}$$

Substituting $(\bar{T}_a - \bar{T}_b) = 4°C$ into Eq. (6.104), one gets

$$\bar{V} = 8 \quad \text{m s}^{-1}$$

The geostrophic wind as estimated from Figs. 1, 8, and 9 of Sweet *et al.* (1981) was about 9 m s^{-1}. Since the geostrophic wind blew at an angle of 45° toward the Gulf Stream, the u component of the geostrophic wind was about $9\cos 45°$, or 6 m s^{-1}. Thus, the added effects of both geostrophic and sea breeze is therefore approximately 14 m s^{-1} or 28 knots on the Gulf Stream side. This is in reasonable agreement with the observation as shown in Fig. 6.22.

Because of the simplicity of the model, there are a number of parameters that must be specified. In order to see how sensitive the $\bar{V}$ solution is to the probable error of estimation of these quantities, the following assessments are made.

(a) Effect of k. Since $k = 2\Omega(\sin \phi)(\tan \psi)$, and for a given latitude, say 37°N, as in our case, ψ may vary from 10° to 30°. Therefore, k may change from 1.6×10^{-5} s^{-1} (for $\psi = 10°$) to 3.2×10^{-5} s^{-1} (for 20°) to 5.1×10^{-5} s^{-1} (for 30°). If one takes the average $\psi \simeq 20°$ over the sea (see, e.g., Roll, 1965), the fluctuation from 10° to 30° will constitute an error of up to 50%, that is, the ratio of 1.6/3.2 between $\psi = 10°$ and 20°.

(b) Effect of $\ln(p_0/p_1)$. The common value of the pressure near the sea surface p_0, is approximately 1012 mb. In offshore regions of the eastern United States, the mean annual morning mixing height is approximately 800 m ($\simeq 920$ mb); and afternoon, 1000 m ($\simeq 900$ mb) (see Figs. 1 and 6 of Holzworth, 1972). Therefore, $\ln(p_0/p_1)$ varies from $\ln(1012/920) = 0.10$ to $\ln(1012/900) = 0.12$, a difference of less than 20%. The reason for using the mixing height in the afternoon is that it represents the maximum vertical convection due to strong latent and sensible heat fluxes. Note that $p_1 = 890$ mb (see preceding section), which is based on Fig. 12a in Sweet *et al.* (1981), is in good agreement with $p_1 = 900$ mb for afternoon mixing height as applied here.

(c) Effect of L. Although the average width of the Gulf Stream is 150 km, at times it may increase to 200 km (Stommel, 1966). Thus the value of M may be reduced by 25% for a given value of $\ln(p_0/p_1)$. The choice of L closely related to the width of the Gulf Stream is necessitated by the fact that as a whole it acts as the heat source for solenoids that drive the circulation.

Since the parameters k, L, and $\ln(p_0/p_1)$ are arranged in the fashion shown in Eq. (6.104), and from discussions above, one may conclude that estimation by Eq. (6.104) is within a factor of 1, with average probable error of no more than 30% if up-to-date weather maps and Gulf Stream charts from satellite are employed.

Sea-breeze-like winds have been observed over the seaward side of the Gulf Stream, which is warm in comparison to the landward side (Sweet *et al.*, 1981). A simple analytical model is proposed to estimate this effect, which is caused by temperature difference across the oceanic front from cold shelf water to warm water. The estimated speed of this sea-breeze-like wind, plus the geostrophic wind effect, is found to be satisfactory for explaining the increase in wind speed from the cold-water side to the warmer-water side of the Gulf Stream. In order to study the detailed interaction between the synoptic and this mesoscale wind system, sophisticated models related to the simulation of the sea breeze (e.g., Estoque, 1962; Pielke, 1974a,b) should be applied and tested. Certainly detailed field experiments to verify the model should be conducted.

Chapter 7 | Air–Sea–Land Interaction

Some atmospheric phenomena in the coastal region are produced directly by the presence of a coastline, which constitutes a major contrast between land and sea in temperature, humidity, wind, and aerodynamic roughness. These systems typically are in micro- and mesoscale dimensions, namely, less than 200 km across the coastal zone.

This chapter deals with some of these phenomena occurring in the coastal zone as a result of air–sea–land interaction, inasmuch as larger-scale systems have been discussed in Chapter 5 and air–sea interactions of smaller scale are also covered in Chapter 6.

7.1 Land- and Sea-Breeze Systems

One of the best examples of air–sea–land interaction is the land- and sea-breeze system. This coastal air-circulation system brings fresh air from the sea in the afternoon to cool coastal residents, whereas farther inland hot and still air is the general rule.

This section deals first with the life cycle of the system, then with simple theoretical considerations, and finally with some cloud pictures taken from meteorological satellites to support deductions reached. For detailed descriptions and numerical modeling for the system, readers are directed to excellent reviews by Atkinson (1981) and Pielke (1984), respectively.

7.1.1 Life Cycle of a Sea-Breeze System

On the basis of various field experiments conducted on the upper Texas coast, Hsu (1970) synthesized an observational model of the coastal air-circulation system as a function of space and time. This system is shown in Fig. 7.1a–d for daytime and in Fig. 7.1e–h for nighttime phenomena. The model was designed with the following considerations in mind:

1. The wind is decomposed into onshore flow and offshore flow. The lower portion of the onshore flow is the sea breeze, and that of the offshore flow is the land breeze.

2. The height or location of each arrow represents the position of the maximum speed (meters per second) of the land or sea breeze.

3. The elliptical shape in the figure represents the extent of the land- and sea-breeze circulation in the locality shown.

4. Since the convective condensation level is approximately at 900 mb, which is about 1 km, and since the 700-mb level (about 3 km) is near the top of the sea-breeze circulation according to observations, these two isobaric surfaces as well as the surface level are labeled in the figure for reference.

5. The horizontal (x, y) plane in the figure shows schematically Galveston Bay and Sabine Lake, the orientation of the coastline, the land and sea areas, and the direction of true north.

We now explain briefly this coastal air-circulation system in sequence as shown in Fig. 7.1a–h. For a more detailed explanation, see Hsu (1969). Let us start at 9 o'clock in the morning (local time 0900 hr in Fig. 7.1a).

(a) At this time, the air temperature over land is still cooler than that over the sea and adjacent waters. Thus, a baroclinic field prevails, and the land breeze is still blowing. Because of the effect of warmer bays and lakes, the land breeze on both ends is weaker than that near the central part of the coastal area, but the land breeze on the right end is relatively stronger than that on the left end if one faces the onshore direction. Since there is a relatively strong land breeze near the central part of the coast between Sabine Lake and Galveston Bay, the return flow over the region 20 km offshore may still produce scattered cumuli at this time, as shown schematically in the figure.

(b) By noon (1200 hr), the land as a whole is warmer than the surrounding waters. A baroclinic field exists, resulting in a direct solenoidal circulation from near shore to about 20 km inland. However, the land breeze still prevails at this time at 20 km inland. Thus, low-level convergence is established. Scattered cumuli with a base at about 1 km tend to form a line. This sea-breeze front passes over a station near the central portion of the coast near late morning. An average sequence of characteristic changes occurs at the surface with the passage of the front: temperature drops approximately 5°F; relative humidity first drops 7%, then rises 14%; and, most pronounced, surface wind direction changes clockwise from northerly to southerly, a total directional change of 180° within 1 hr.

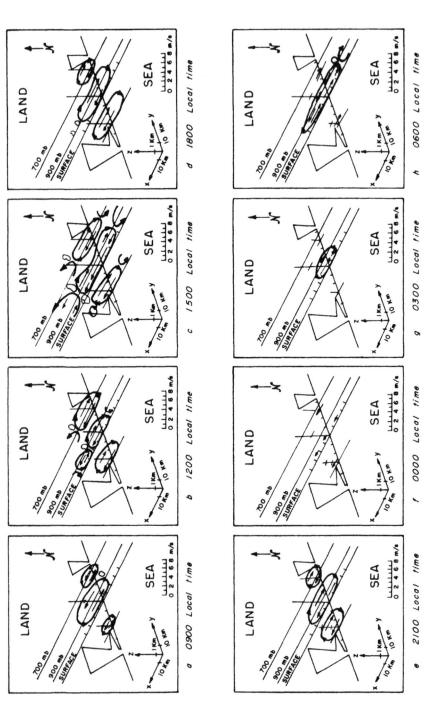

Fig. 7.1. A synthesized observational model of the coastal air-circulation system over the upper Texas coast (Hsu, 1970).

(c) At 3 o'clock in the afternoon (1500 hr), the sea breeze is in the fully developed stage, since the air temperature difference between land and water reaches maximum near noon. Superadiabatic lapse rates appear near the surface on land. At this time, there is a definite chance of formation of cumulus clouds, and showers will reach the ground at 4 to 5 o'clock in the afternoon 30–40 km inland. The convergence line is fully developed at this time also, and the orientation of this line is approximately parallel to the coastline. Because of velocity divergence and relatively dry air in the return flow of the sea breeze, there is a subsidence phenomenon near the coastal area.

(d) About 6 o'clock in the afternoon (1800 hr), the land as a whole is still warmer than surrounding waters, but the baroclinic field is weaker than it was about 3 hr earlier. At this time, the sea breeze and its return flow are still prevailing. A few cumuli may still be lingering in the sky from 30 km and farther inland. Because the bay on the left is bigger than that on the right, the sea breeze at this time is stronger on the left end than on the right (if one faces the onshore direction).

(e) At 9 o'clock in the evening (2100 hr, Fig. 7.1e), the sun has set. At this time, the residual sea breeze and its return flow may be still prevailing, but the speed and strength are weaker than they were at 6 p.m., particularly near the surface.

(f) Near midnight (2400 hr), the cool pool phenomenon caused by the effect of the sea breeze plus nocturnal cooling is about to form. The central portion of the pool is roughly parallel to the shoreline about 20 km inland. At this time, a temperature inversion and occasionally fog appear over land. The nearly barotropic field is prevailing over the area. The sea breeze at this time is almost gone. Surface wind is nearly calm on land.

(g) About 3 o'clock in the morning (0300 hr), the cool pool is formed more completely. However, because the bays and lakes are warmer at this time, the horizontal temperature gradient from land to sea is larger near the central portion of the coast than at both ends in the area; thus, the land breeze will start earlier near the central portion of the coast. Since the cool pool is analogous to a cool "high" presssure region, a stable and barotropic situation will exist beyond 10 km inland. In the nearshore area, however, owing to the cool pool and the warmer seawater, a baroclinic field exists from the vicinity of the shoreline to 20 km offshore.

(h) About 6 o'clock in the morning (0600 hr), the horizontal temperature gradient from land to sea is at its maximum. Thus, the land breeze will be stronger than at 0300 hr. A weak land-breeze convergence line associated with more scattered cumuliform clouds may be observed about 30 km offshore. The land breeze will continue to blow to the midmorning hours, and the sea-breeze cycle will start over again, as shown in Fig. 7.1a. Thus the coastal air-circulation system has completed its life cycle.

7.1.2   A Simple Sea-Breeze Model

The sea-breeze circulation model to be considered here requires a temperature difference between the air over a coastal plain ($\bar{T}_a$) and that over water ($\bar{T}_b$), that is, $\bar{T}_a - \bar{T}_b$. According to the customary form of Bjerknes's circulation theorem (see, e.g., Hess, 1959), if friction is neglected, the intensity of the sea breeze will increase until values of $\bar{T}_a - \bar{T}_b$ change from positive to negative. According to Hauwritz (1947), however, owing to the effect of friction, maximum sea-breeze intensity occurs not when $\bar{T}_a - \bar{T}_b$ has decreased to zero but earlier, while the land is still warmer than the sea. The reason for this is that a specific, positive temperature difference is required simply to overcome the frictional force. Direct solenoidal circulation causes progressively increasing wind speeds in the sea breeze during the day, while surface friction exerts a braking effect. Thus a balance is possible between the increase in circulation caused by the mass distribution and the decrease caused by retarding friction; consequently, steady winds are theoretically possible.

Let us consider a sea-breeze model that takes friction into account. We assume, as did Hauwritz, that the circulation takes place in a vertical x, z plane, with the x axis perpendicular to the coastline. The path of integration is shown in Fig. 7.2, where $\bar{T}_a$ and $\bar{T}_b$ represent mean temperatures between the two pressure levels. The Coriolis force is disregarded, since it does not affect our main conclusions (see Chapter 6). The frictional force is assumed to be opposite to, and proportional to, the wind velocity. With these assumptions, the equations of motion can be written in the following form:

$$\frac{du}{dt} + ku = -\frac{1}{\rho}\left(\frac{\partial p}{\partial x}\right) \tag{7.1}$$

$$\frac{dw}{dt} + kw = -\frac{1}{\rho}\left(\frac{\partial p}{\partial z}\right) - g \tag{7.2}$$

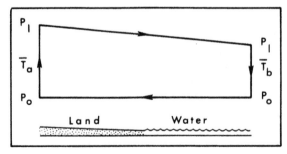

Fig. 7.2. Path of integration at the time of onset of sea-breeze circulation.

where u and w are the velocity components in x and z directions, respectively, p is the pressure, ρ is the density, g is the acceleration of gravity, and k is a constant that expresses the intensity of the frictional force. The resulting circulation may be expressed as

$$C = \oint (u\,dx + w\,dz) = L\bar{V} \tag{7.3}$$

where L is the length of the path of integration and $\bar{V}$ is the mean speed of the sea-breeze circulation along the path of integration.

By multiplying Eq. (7.1) by dx, Eq. (7.2) by dz, adding, and integrating around the closed path, one derives the rate of change of the circulation C to be

$$\frac{dC}{dt} = \oint \left(\frac{du}{dt}\,dx + \frac{dw}{dt}\,dz \right) = -\oint \frac{dp}{\rho} - \oint g\,dz - kC \tag{7.4}$$

The second integral of Eq. (7.4) is zero because it is the closed line integral of an exact differential, assuming g to be a single-valued function of z. From the chosen path of integration, Eq. (7.4) can be written in the form

$$dC/dt = R(\bar{T}_a - \bar{T}_b)\ln(p_0/p_1) - kC \tag{7.5}$$

where R ($= 2.87 \times 10^6$ erg g^{-1} K^{-1}) is the gas constant for air. Then, from Eqs. (7.3) and (7.5), one gets

$$\frac{d\bar{V}}{dt} + k\bar{V} = (\bar{T}_a - \bar{T}_b)\frac{R}{L}\ln\frac{p_0}{p_1} = M(\bar{T}_a - \bar{T}_b) \tag{7.6}$$

The quantity $M = (R/L)\ln(p_0/p_1)$ is constant, since for a fixed path of integration p_0 and p_1 are constant.

In the sea-breeze problem, $\bar{T}_a - \bar{T}_b$ is assumed to be a periodic function of time, as observed by Hsu (1967), so that

$$M(\bar{T}_a - \bar{T}_b) = A\cos\omega t \tag{7.7}$$

where ω ($= 7.29 \times 10^{-5}$ s^{-1}) is the angular velocity of the earth's rotation. This assumption implies that time is reckoned from the instant when $\bar{T}_a - \bar{T}_b$ reaches its maximum, that is, when $t = 0$, then $\bar{T}_a - \bar{T}_b$ has its maximum value. Putting Eq. (7.7) into Eq. (7.6) and integrating, we find the solution of the resulting differential equation to be

$$\bar{V} = \text{const}\,e^{-kt} + A(k^2 + \omega^2)^{-1}(\omega\sin\omega t + k\cos\omega t) \tag{7.8}$$

Thus, in the sea-breeze area, if the temperature difference between the air over land and that over water and the path of integration are known, Eq. (7.9) may be used to calculate the mean speed of the sea-breeze circulation.

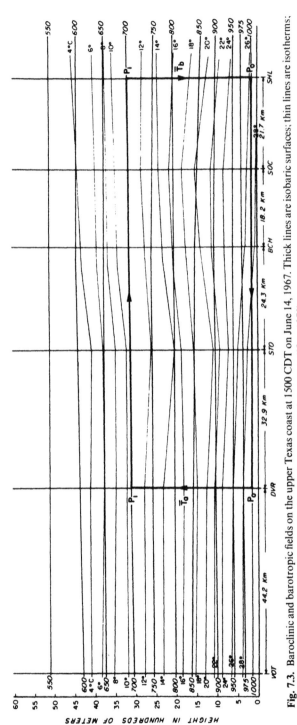

Fig. 7.3. Baroclinic and barotropic fields on the upper Texas coast at 1500 CDT on June 14, 1967. Thick lines are isobaric surfaces; thin lines are isotherms; and the heavy line is the path of integration for the sea-breeze circulation system (Hsu, 1970).

The arbitrary constant in Eq. (7.8) can be assumed to be zero because, in the absence of a temperature difference ($A = 0$), the wind should be zero. Hence, Eq. (7.8) becomes

$$\bar{V} = A(k^2 + \omega^2)^{-1}(\omega \sin \omega t + k \cos \omega t) \tag{7.9}$$

In the Texas coast sea-breeze area, the path of integration may be determined from the observational data, as shown in Fig. 7.3, that is, $P_0 = 1000$ mb, $P_1 = 700$ mb, $L = 200$ km, and $\bar{T}_a - \bar{T}_b = 5°C$. This is slightly larger than the circulation area considered by Fisher (1960). Substituting these values into Eqs. (7.6) and (7.7), one finds that

$$K = 0.05 \quad \text{cm s}^{-2} \text{ deg}^{-1} \quad \text{and} \quad A = 0.25 \quad \text{cm s}^{-2}$$

According to observations (Hsu, 1967, Fig. 32), $\bar{T}_a - \bar{T}_b$ reaches its maximum around 1200 CST (that is, $t = 0$ at 1200 CST). If we assume the friction $k = 2 \times 10^{-5}$ s^{-1} in the coastal area, as suggested by Hauwritz (1947), and substitute $t = 0$ into Eq. (7.9), we find that the mean speed of the sea-breeze circulation ($\bar{V}$) perpendicular to the Texas coast is 8.8 m s^{-1}. This is in excellent agreement with the observations of 8.0 m s^{-1} shown in Fig. 7.1 at the time of maximum sea breeze.

7.1.3 Examples of Sea-Breeze Circulation from Satellite Photography

Examples of sea-breeze systems are shown in Figs. 7.4 and 7.5. More examples are given in Fett (1979). Note that in coastal regions where there are mountains, such as near the east end of Cuba in the area of Guantanamo Bay, the combination of sea breeze and upslope wind will produce more convective clouds as shown in Fig. 7.5.

7.2 Coastal Upwelling

Upwelling is a slow upward motion of deeper water. Upwellings are important in coastal meteorology because coastal stratus and fog usually form over these cool waters. As shown in Fig. 7.6, coastal upwelling is caused by a wind stress acting along the boundary of an ocean in combination with the effect of the earth's rotation (Coriolis effect) to produce a net mass transport of water offshore in the surface layer. These offshore-flowing waters are replaced by cold water being upwelled from subsurface layers (Fig. 7.7).

Mathematically the upwelling phenomenon may be explained by the Ekman spiral as discussed in the previous chapter. For oceanography,

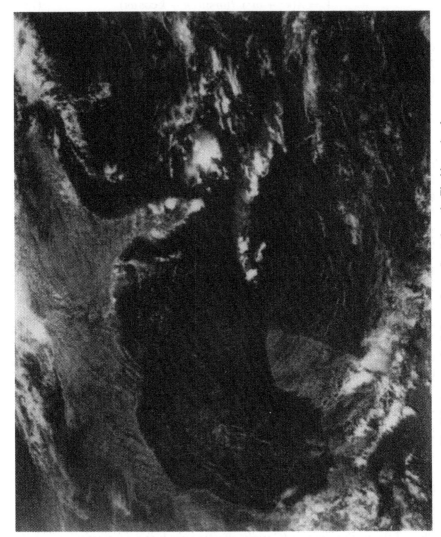

Fig. 7.4. An example of the sea-breeze front along the Florida peninsula.

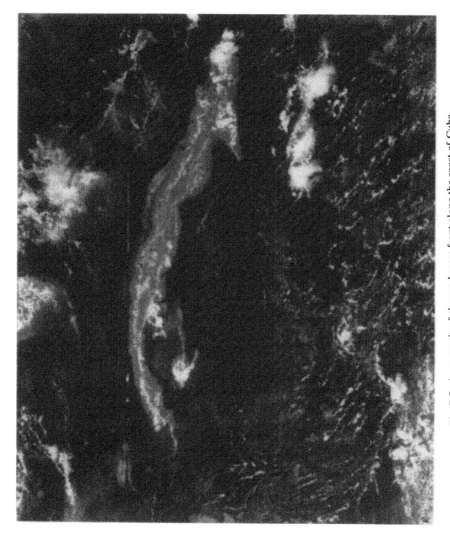

Fig. 7.5. An example of the sea-breeze front along the coast of Cuba.

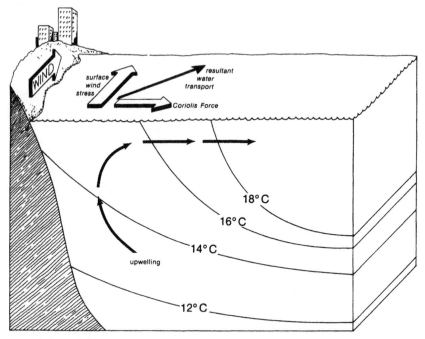

Fig. 7.6. Schematic representation of water circulation in coastal upwelling in response to equatorward wind stress (northern hemisphere). [After Fett (1979).]

however, the following solution can be derived (see, e.g., Neumann and Pierson, 1966, pp. 192–193):

$$u = V_0 e^{-(\pi/D)z} \cos\left(45° - \frac{\pi}{D}Z\right)$$

$$v = V_0 e^{-(\pi/D)z} \sin\left(45° - \frac{\pi}{D}Z\right)$$

(7.10)

where

$$D = \pi\left(\frac{2A}{\rho_{sea}f}\right)^{1/2}$$

(7.11)

and

$$V_0 = \frac{\tau_y}{(\rho_{sea}Af)^{1/2}}$$

(7.12)

in which A is an eddy viscosity coefficient, f is the Coriolis parameter, and $\tau_y (= \rho_{air}C_D U^2)$ is the wind stress (cf. Chapter 6) in the y direction.

Fig. 7.7. An example of a satellite detection of coastal upwelling along the cloud-free region offshore of California at 0409 GMT on August 30, 1977. Cold water is white, and warmer water is dark. [After Fett (1979).].

Equation (7.10) shows that at the sea surface (i.e., $Z = 0$), $u = V_0 \cos 45°$ and $v = V_0 \sin 45°$. Hence, the surface current vector V_0 moves in a $45°$ angle to the right of the wind direction in the northern hemisphere, as shown in Fig. 7.6, and $45°$ to the left in the southern hemisphere. Since the upwelled water comes to the surface from depths that range from 50 to 300 m, D may be assigned a value of, say, 50 m. If a location is given, f is known. Then A may be computed from Eq. (7.11). From A and wind speed (thus τ_y), V_0 can be estimated from Eq. (7.12). An example of this computation of an upwelling event offshore from California is given in Bishop (1984).

Equation (7.12) also indicates that for a given depth and location (thus A and f are constant), V_0 is directly proportional to τ. The wind stress, which was discussed in detail in the previous chapter, depends in turn on wind speed, duration, fetch, and direction. Therefore, coastal upwelling has seasonal variability.

Coastal upwelling is most pronounced off the western U.S. and North Africa coasts from April through August, and off the Peru and Chile coasts from March through May (Fett, 1979). Although the upwelling season lasts several months, winds do not blow steadily along the coast, producing upwelling, during the whole period. Instead, short-term "upwelling events" occur on a time scale of days or weeks. The upwelling events are characterized by an increase in the longshore component of the wind and, since the sea-surface temperature field responds rapidly to changes in wind forcing, a rapid drop in the sea-surface temperature occurs near the coast (Holladay and O'Brien, 1975).

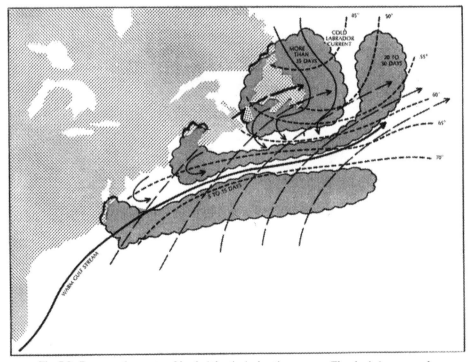

Fig. 7.8. Fog over the western North Atlantic during the summer. The shaded areas are fog; numbers in the middle refer to the average number of days with fog during the period June–August. Dashed lines are sea-surface temperature (isotherms) in degrees Fahrenheit. Dashed arrows are streamlines of the prevailing surface air currents. Solid arrows represent the cold Labrador Current and the warm Gulf Stream. [After Kotsch (1983).]

7.3 Coastal Fog

Fog is a visible aggregate of many minute water droplets that are suspended in the atmosphere near the earth's surface. Fog differs from cloud
only in that the base of fog is at the earth's surface, while clouds are
above the surface (Huschke, 1959). Examples of coastal fogs are shown in
Figs. 7.8–7.11.

There are many kinds of fog, but advection fog is the most prevalent in the
coastal zone. According to Kotsch (1983), when warmer air blows over a cold
land or water surface, it gives off heat to the colder underlying surface. This
process will cool the air to the dew point and advection fog may form. Over
land, if the wind speed is higher than, say, 10 m s^{-1}, the air will be mixed
through a relatively deep layer because of larger frictional effect. Thus low
stratus or stratocumulus clouds, rather than fog, will form. Over the sea,
however, because there is less frictional effect than over land, advection fog will
form even when wind speeds are 30 knots (15 m s^{-1}). It is usually extensive and
persistent.

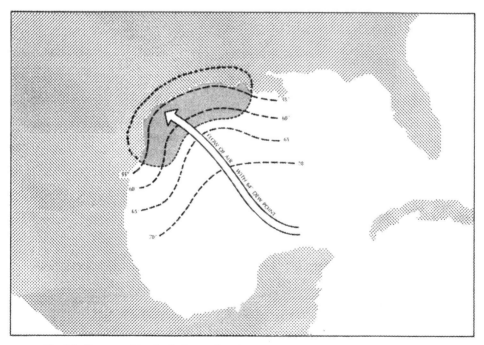

Fig. 7.9. Fog caused by the drainage of cold river water into the Gulf of Mexico during the
winter. Shaded area is fog. Dashed lines are sea-surface temperature lines (isotherms) in degrees
Fahrenheit. [After Kotsch (1983).]

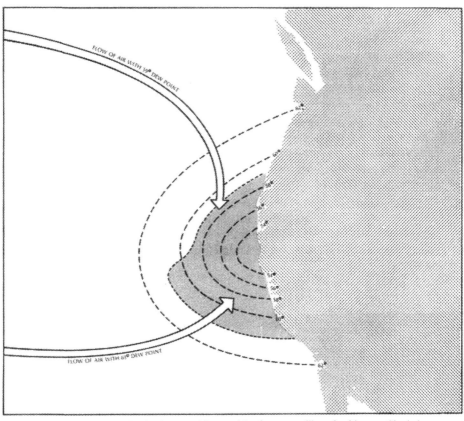

Fig. 7.10. Northern California coastal fog resulting from upwelling of cold water. Shaded area is fog. Dashed lines are sea-surface temperature lines (isotherms) in degrees Fahrenheit. [After Kotsch (1983).]

In coastal regions, advection fogs form mainly when winds having high moisture content blow from the southerly direction over cold-water regions. Examples are shown in Fig. 7.8, where the cold-water Labrador Current crosses the North Atlantic. Over scattered parts of the northern section of the Gulf of Mexico, mainly during the winter season, the drainage of cold, fresh water from the Mississippi River and coastal bays and estuaries into the Gulf will cause fog to form when water temperatures are lower than the dew point of the warm air, as shown in Fig. 7.9. Figure 7.10 shows fog over cold coastal upwelling waters off the coast of California. Because of contrasting water temperatures, such as along the Aleutian Island chain, where the Bering Sea is colder than the North Pacific, fogs may persist (Fig. 7.11).

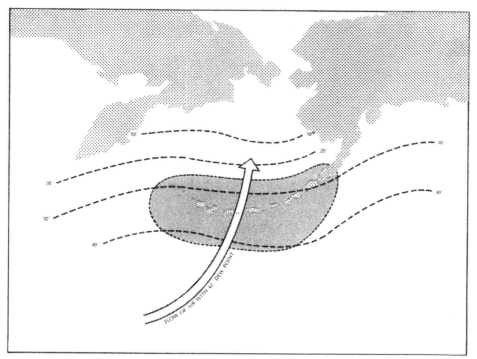

Fig. 7.11. The famous Aleutian fog, which persists even when the winds reach 35 knots. The shaded area is fog. Dashed lines are sea-surface temperature lines (isotherms) in degrees Fahrenheit. [After Kotsch (1983).]

7.4 Coastal Jets

According to Kraus *et al.* (1985), no unique definition for the term "low-level jet" exists in the literature. This is mainly because of the fact that low-level wind maxima in the boundary layer occur in a number of rather different situations: with strong synoptic-scale baroclinity; with fronts; in cases of advective accelerations; in connection with mountain barriers if flow splitting and confluence occur; in land and sea breezes; in mountain and valley winds; and finally in ageostrophic boundary-layer flow, when stable stratification reduces wind speeds (Blackadar, 1960).

Some examples of low-level jets (LLJ) existing in coastal regions are described in this section.

Over flat, open coasts such as along the U.S. Gulf of Mexico from Texas to Florida, conditions are sometimes right for maximum winds to develop at night. Two types of these jetlike winds have been observed (Hsu, 1979a). One is

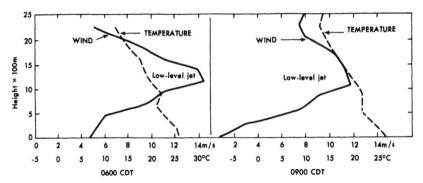

Fig. 7.12. An example of a low-level jet that blew over the beach (left diagram) and over a site about 24 km inland from the Gulf of Mexico (right diagram) between Sabine Pass and Galveston, Texas, on June 14, 1967. Note that the jet was situated at the top of the elevated inversion layer (Hsu, 1979a). Copyright © by D. Reidel Publishing Co. Reprinted by permission.

the familiar synoptic-scale, low-level jet over the coast and Great Plains (see, e.g., Lettau, 1954; Blackadar, 1960). Figure 7.12 shows an example of a low-level jet over the coast near Galveston, Texas (Hsu, 1979a). The other type, which may be less well known, is the mesoscale wind maximum. Unlike the low-level jet, which occurs at the top of the nocturnal inversion and only with a wind from the southerly quadrant (Blackadar, 1960), the mesoscale jet may exist between two inversion layers: the mesoscale elevated inversion and the microscale (nocturnal) ground-based inversion just above the "cool pool," or "mesohigh," on the coast (Hsu, 1970). It can occur when the wind blows from other quadrants as long as it is not parallel to the shoreline. The vertical variations of these jetlike winds are shown in Fig. 7.13. The winds were measured mainly in February 1977 by radiosondes, which were tracked by theodolite. The site was about 18 km from the Gulf near Tarpon Springs, northwest of Tampa, Florida. From February 7 through 12, 1977 the study area was dominated by high pressure; light geostrophic winds blew from land to sea, and fair weather prevailed. Temperature and wind measurements shown in Fig. 7.13 from February 7 through 9 indicated that during the night and early morning, jetlike winds occurred between 100 and 600 m. Unlike the familiar low-level jet, which occurs at the top of the nocturnal inversion, these jetlike winds existed generally above the lowest inversion layer but within or beneath the inversion layers at higher altitude. During the day such jetlike winds diminished.

The spatial distribution of this jetlike wind is shown in Fig. 7.14. Atmospheric sounding about 50 km offshore was made by a tethered balloon sounding system from a boat. Station Ruskin is a National Weather Service

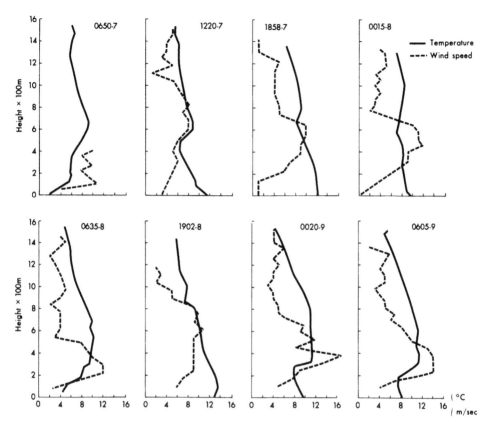

Fig. 7.13. Examples of the jetlike winds within the planetary boundary layer over a site about 18 km inland from the Gulf of Mexico near Tarpon Springs, northwest of Tampa, Florida. In the figure, numbers stand for the time and the day in February 1977. Note that, unlike that in Fig. 7.12, this type of nocturnal jet was situated between surface-based and elevated inversion layers (Hsu, 1979a). Copyright © by D. Reidel Publishing Co. Reprinted by permission.

rawinsonde station that is about 65 km from the main Gulf coastline. Tarpon Springs is located about 18 km inland from the shoreline. Anclote Power Plant is near the shoreline, and Anclote Island is about 11 km offshore. Note that the rawinsonde observation at Ruskin was made at 0600 on February 11, while the others were made between 2340 on February 10 and 0020 on February 11. In Fig. 7.14 it is shown that around midnight the jet had a maximum speed of 15–16 m s^{-1} at a height between 300 and 500 m above sea level near the coast. The jet was located above the inversion layer, and its directional shear was about 40°. Furthermore, the air temperature offshore was on an average 5°C warmer than that over the land.

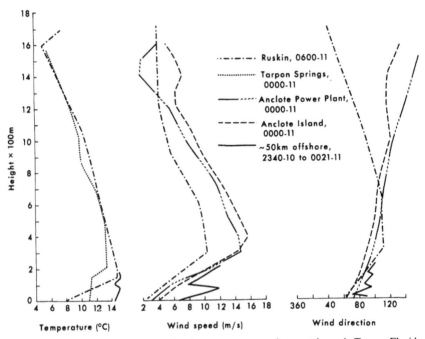

Fig. 7.14. Spatial variability of the jet phenomenon across the coastal zone in Tampa, Florida, region (see text for explanation) (Hsu, 1979). Copyright © by D. Reidel Publishing Co. Reprinted by permission.

It has been shown in Hsu (1979a) that the necessary conditions for the existence of such a phenomenon are as follows:

(a) The prevailing geostrophic wind must be light, preferably less than $5–6 \text{ m s}^{-1}$.

(b) The said geostrophic wind preferably will blow from land to sea.

(c) The change in wind direction with height in the planetary boundary layer should be less than $80°$.

(d) From a few simultaneous radiosondings both offshore and onshore, it is also clear that the air temperature over adjacent seas should be at least $5°C$ warmer than that over the land during the night, so that the "cool pool" or mesohigh over the land can form (Hsu, 1970).

On the basis of the above conditions, a possible mechanism to explain such a phenomenon is illustrated in Fig. 7.15. For simplicity, only offshore wind conditions are given. Figure 7.15 shows that at times five regions may exist over a flat, open coast in which this phenomenon may occur. They are produced by the existence and variation of two distinct inversion layers: the mesoscale inversion layer and the nocturnal microscale inversion height.

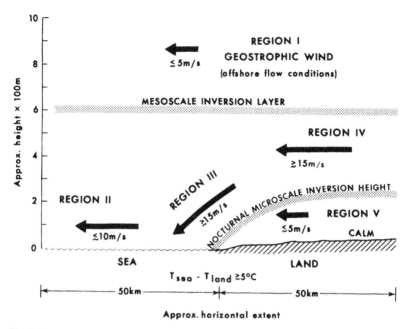

Fig. 7.15. A possible mechanism to explain the jetlike winds occurring in region IV (see text for details) (Hsu, 1979). Copyright © by D. Reidel Publishing Co. Reprinted by permission.

Nocturnal inversion layers are formed primarily as a result of radiational cooling near a rather uniform land surface. This "cool pool" or mesohigh has been documented by Hsu (1970) on the upper Texas coast and is shown also in Fig. 7.13 for the Gulf Coast of central Florida. Note that near-calm wind conditions usually exist within this "cool pool," particularly near the ground.

When these two inversion layers exist, Venturi and gravity-wind (or drainage or katabatic wind) effects are pronounced in turn. The relative wind speeds in Fig. 7.15 are for illustration and comparison only. Particular attention should, however, be given to region III, where Venturi and gravity-wind effects are combined, and to region IV, where the Venturi effect is pronounced. These effects are the possible mechanism for the generation of observed jetlike winds that exist principally between the inversion layers produced by microscale effects near the bottom and mesoscale effects on the top. This high-speed shear zone occurs between 100 and 600 m above the area a few kilometers offshore to at least 40–50 km inland. Over the adjacent sea the jet may be within 100 m above the sea surface.

An estimate of the Venturi effect may be made as follows: Assuming the two-dimensional structure is valid as shown in Fig. 7.15, we have

$$H_{II}U_{II} = H_{IV}U_{IV}$$

where H and U are the height and wind speed, respectively, and subscripts II and IV represent regions II and IV.

Since H_{II} is about 600 m, U_{II} is about 7 m s^{-1} (averaged between 10 m s^{-1} at about 100 m height over the sea surface and the geostrophic wind of 5 m s^{-1} at the top near the mesoscale inversion layer), and H_{IV} is about 300 m. The term U_{IV} is estimated at approximately 14 m s^{-1}, which is in agreement with the observed values of around 15 m s^{-1} (Figs. 7.13 and 7.14). The wind in region IV is directed down the slope of the incline along the nocturnal microscale inversion height (Fig. 7.15). The greater air density near the slope also causes region III to experience higher wind speeds.

A nocturnal low-level jet characterized by a distinct inertial oscillation lasting from around sunset until sunrise the next day was observed during a coastal experiment in Germany (Kraus *et al.*, 1985). Figure 7.16 shows profiles of wind speed at four stations. One station is Bremervörde, 50 km inland from the coast, where the wind is measured by radar-tracked pibals. The other three stations form a triangle of approximately 20 km side length with center approximately 30 km southeast of Bremervörde; at these stations wind was measured by tethersondes using cup anemometers and compasses. The profiles show a jet maximum with good agreement among the four independently measured profiles during the night. However, in the morning, when turbulence leads to larger fluctuations and thus dissolves the jet, the profiles differ considerably. The observed nighttime consistency shows that this LLJ developed as a mesoscale phenomenon in an area larger than 40 km in diameter.

Low-level wind maxima were observed at Näsudden, on the southern coast of Gotland Island, Sweden. As reported by Högström and Smedman-Högström (1984):

> During a period of several days at the end of May 1980 the synoptic situation over the Baltic Sea was characterized by a very steady southeasterly flow (see Fig. 7.17). As the sea surface was appreciably colder than the approach flow—the sea surface temperature was $\sim 6°C$ and the daytime air temperature near the Latvian coast was above 20°C—the lowest layers of the atmosphere over the Baltic Sea attained a strongly stable stratification. This situation brought about a remarkable wind regime at Näsudden, characterized by a strong low-level jet present most of the time.

Figure 7.18a shows that there is a very strong inversion in the lowest few hundred meters during the night. During the day a shallow, ~ 90-m-deep convective boundary layer is observed at Näsudden, but over the sea the inversion is no doubt ground-based the entire time. Figure 7.18b shows the detailed structure of wind field throughout the lowest 2000 m during the period from May 28 at 1400 LST to May 29 at 0700 LST. It can be seen that there is a strong wind gradient in the lowest 200 m, and at times there is a pronounced low-level jet present at about that height.

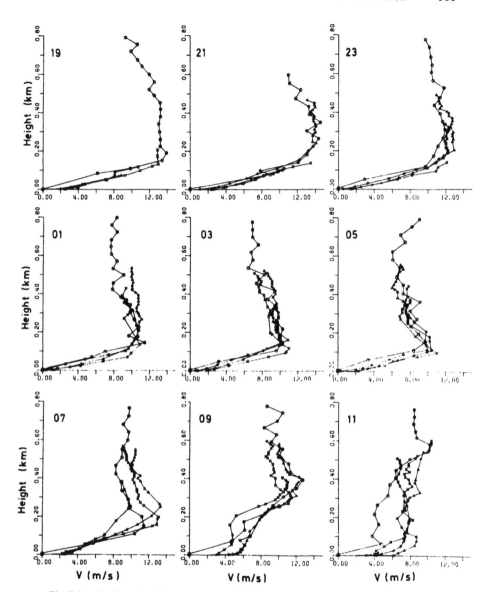

Fig. 7.16. Profiles of wind speed V for the station Bremervörde (○) and the tethersonde triangle Stemmen (▲)–Oldendorf (∗)–Ahrenswohlde (☒) for the night from September 30 to October 1, 1981. The launching time in GMT of the pibal and tethersonde systems is given in the upper left corner of each diagram. [After Kraus *et al.* (1985). Copyright © by D. Reidel Publishing Co. Reprinted by permission.]

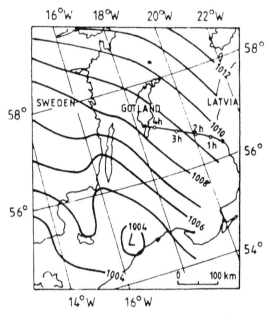

Fig. 7.17. Map showing the area of the low-level jet case study with Gotland in the middle of the Baltic Sea. The isobars are for May 29, 1980, at 0100 hr. Also included is a trajectory from the Latvian coast to Näsudden (the cross) at Gotland. [After Högström and Smedman-Högström (1984). Copyright © by D. Reidel Publishing Co. Reprinted by permission.]

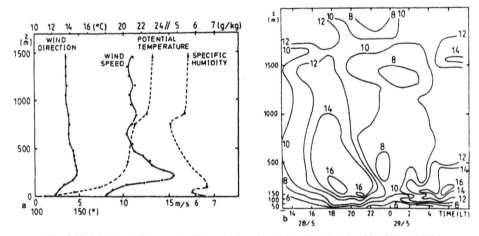

Fig. 7.18. (a) Pilot balloon and radiosonde data from Näsudden for May 29, 1980, at 0500 hr, showing wind direction, wind speed, potential temperature, and specific humidity. (b) Wind-speed cross section at Näsudden for the period May 28 at 1400 hr to May 30 at 1200 hr. Below 145 m wind data are taken from anemometer measurements in a mast. Above that height the wind data were obtained from a series of double theodolite pilot balloon measurements. [After Högström and Smedman-Högström (1984). Copyright © by D. Reidel Publishing Co. Reprinted by permission.]

7.5   Other Coastal Phenomena

There are many other mesoscale meteorological phenomena that occur in various coastal regions. Descriptions of some of these phenomena over land can be found in Yoshino (1975, 1976). For discussion of such phenomena over coastal waters, refer to the "Mariners Log," published by the U.S. National Oceanic and Atmospheric Administration. Because of space limitations, only a few topics are illustrated in this section.

7.5.1   Local Winds in Southern Greenland

The following material is excerpted from Mertins (1976).

Near coasts, and especially near mountainous coasts, local wind effects occur. Figure 7.19 depicts the weather situation around southern Greenland on January 9, 1953, at 0000 GMT. A north-northeast wind of full hurricane force, with poor visibility due to snow and spray, was blowing near Cape

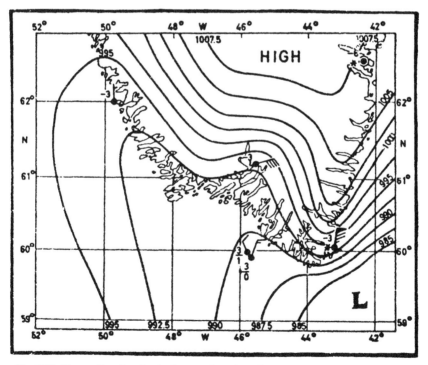

Fig. 7.19. The weather situation in the area of southern Greenland on January 9, 1953, at 0000 GMT. [After Mertins (1976).]

Farewell, but some 60 miles to the west two trawlers experienced good fishing weather with good visibility and wind of no more than force 3–4 (less than 10 m s^{-1}).

Such local contrasts occur when an easterly or northeasterly wind of sufficient force blows against the Greenland Icecap. This causes "blocking action" to take place along the east coast, and the blockage results in a ridge of high pressure over eastern Greenland with a very steep pressure gradient in the open sea to the southeast. By contrast, a trough of low pressure with a very weak pressure gradient develops along the southwestern coast.

Such contrasting conditions between the coastal waters to the east and to the west of Cape Farewell may be anticipated by observing storm tracks on a weather map. For instance, when a depression between about 44°N and 30°W heads north, this type of situation is likely to develop, the high pressure over central Greenland and the Denmark Strait area favoring its development. In these circumstances, fishing trawlers usually sail to the west, while any ships sailing from West Greenland bound for Europe may round Cape Farewell farther south than usual unless they prefer to wait for the weather to improve. Figure 7.20 illustrates this point. The trawler *Hanseat* took a southerly course, rounding Cape Farewell at about 57°N. It experienced northeasterly gales

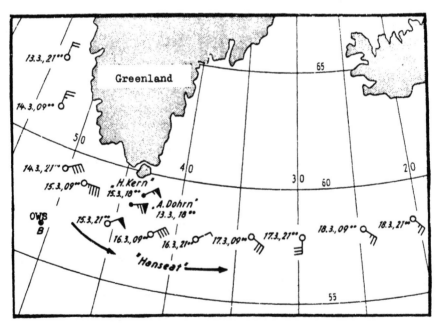

Fig. 7.20. Observations of wind from the trawler *Hanseat* at 0900 and 2100 GMT in March 1962. [After Mertins (1976).]

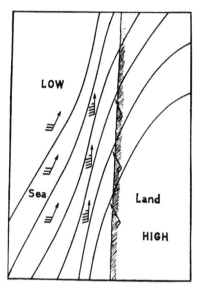

Fig. 7.21. A model of the wind distribution along a mountainous coast. [After Mertins (1976).]

(force 8–10; > 20 m s^{-1}), while other ships farther to the north nearer Cape Farewell experienced northeasterly winds of hurricane force (> 30 m s^{-1}).

Another example is illustrated by Mertins (1976) in Fig. 7.21, which shows the wind pattern along a mountainous coast with the isobars cutting across the coastline at an acute angle. The airflow is blocked along the coast and consequently a zone with a steeper pressure gradient is established between the coastal region and the undisturbed pressure field farther out at sea. This results in a strong wind zone parallel to the shore, which ships can avoid either by seeking more sheltered conditions near the coast or, preferably, by sailing farther out to sea.

7.5.2   An Orographically Induced Mesoscale Cyclone

The following is drawn from Reed (1980).

Sustained winds in excess of 70 knots and gusts of close to 100 knots were measured at the Hood Canal Bridge in western Washington on the morning of February 13, 1979, shortly before the bridge collapsed. An extraordinary blowdown of timber also occurred in the nearby area. Elsewhere in western Washington, maximum 1-min wind speeds were generally 40 knots or less and peak gusts were mostly under 65 knots. Wind damage was widespread but less concentrated than in the vicinity of the bridge.

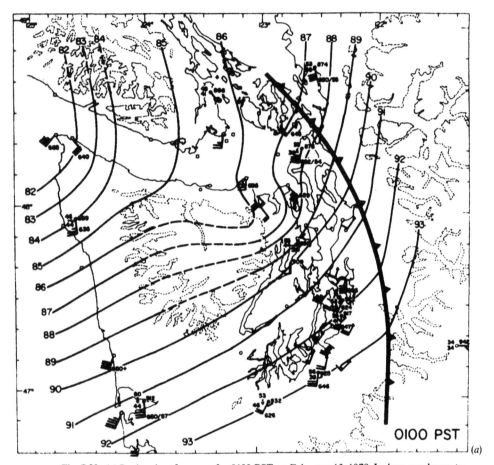

Fig. 7.22. (a) Regional surface map for 0100 PST on February 13, 1979. Isobars are drawn to 1-mb intervals (dashed at higher elevations). Winds are in knots. Maximum gusts (G) at observation time and in preceding hour are plotted below stations. [After Reed (1980).] (b) Regional surface map for 0400 PST on February 13, 1979. (c) Regional surface map for 0700 PST on February 13, 1979. (d) Regional surface map for 1000 PST on February 13, 1979.

For this orographically induced mesoscale cyclone, the meteorological conditions are shown in Fig. 7.22a–d (Reed, 1980).

At 0100 PST on February 13 (Fig. 7.22a) a cold front had just passed through the Puget Sound region. Wind directions behind the front ranged from southeast to southwest, depending on the local topography. The highest speeds (45 knots) were observed at coastal stations. Inland a 40-knot wind was recorded on the Evergreen Point Bridge, and 35-knot velocities were measured at a number of locations, including the Coast Guard cutter *Campbell*

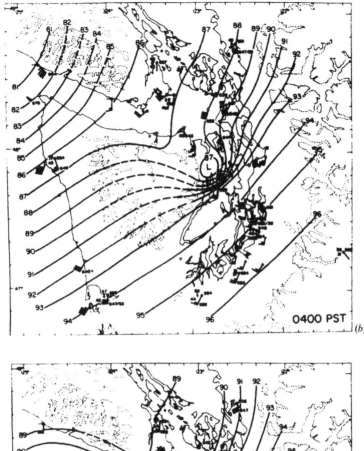

0400 PST

(b)

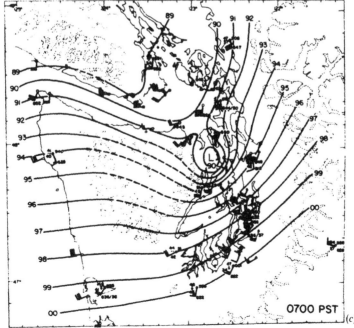

0700 PST

(c)

Fig. 7.22. *(continued)*

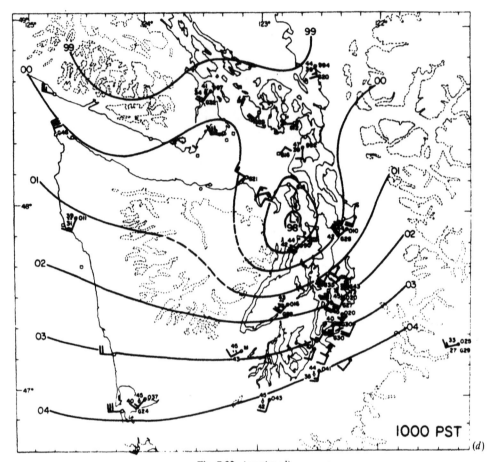

Fig. 7.22. (*continued*)

(CG). The winds in the area about Sequim Bay gave evidence of a trough formation in the lee of the Olympics.

The isobaric analysis for 0400 PST (Fig. 7.22b) indicates a further strengthening of the lee trough and the possible formation of a low-pressure center in the area between the Hood Canal Bridge and Sequim Bay. In connection with the formation of the low, a remarkable pressure gradient developed between the *Campbell* (CG) and Bangor (KB). The pressure difference of 5 mb resulted from a combination of a continued small lowering of the pressure in the vicinity of the bridge and a moderate rise at Bangor. The rise at Bangor was part of the general increase of pressure that took place behind the cold front. To assure that the barometric readings at the two sites were comparable, the readings for both were corrected on the basis of comparisons with readings at nearby stations before the storm.

With the formation of the extreme pressure gradient, the wind at the *Campbell* had increased to nearly 60 knots by the 0400 PST observation time. According to the watch officer, the rise was quite abrupt, coming shortly after 0300 PST, when a band of dark clouds advanced from the south and winds, which previously had been averaging 30–35 knots, jumped by about 25 knots and held at the new level. Winds elsewhere were not as extreme, though still appreciable. Several stations reported speeds above 40 knots, and Cape Flattery (CF) recorded its peak sustained wind of nearly 50 knots. The cause of the steplike rise in the wind force at the *Campbell* is not apparent. If the same phenomenon occurred at the control tower on the bridge, it was unnoticed by the bridge tenders.

The map for 0700 PST (Fig. 7.22c), the time of the bridge collapse, provides further evidence for the existence of a mesoscale low or vortex, the wind at the southern tip of Sequim Bay having shifted to light northerly. The *Campbell* at that hour was taking shelter in lower Dabob Bay, experiencing winds of only 20 knots. Residents at the head of the bay, however, reported much stronger, damaging winds. The barometric pressure on the *Campbell* was now in close agreement with that at Bangor. The strongest wind seen on the map at this hour is 60 knots at the Evergreen Point Bridge. At the Seattle Urban station (SU), only 2 miles to the west, the speed is 15 knots, a striking illustration of the effect of exposure to wind behavior. The latter site is in a hollow sheltered by a hill to the south. At the bridge, southerly winds have an overwater fetch of nearly 5 miles.

By 1000 PST (Fig. 7.22d) winds had subsided everywhere, and the pressure gradient had diminished greatly. The wind and pressure reports at that hour clearly substantiate the existence of a mesoscale vortex between lower Hood Canal and Sequim Bay.

7.5.3   Gap Winds in the Strait of Juan de Fuca

According to Overland and Walter (1981), gap winds can be defined as a flow of air in a sea-level channel that accelerates under the influence of a pressure gradient parallel to the axis of the channel. Figure 7.23 shows an example of gap winds observed in the Strait of Juan de Fuca, between western Washington State and British Columbia. There was a high-pressure region over central British Columbia and a low-pressure system propagating northward, seaward of the Washington coast, thus producing strong easterly winds of $13–15$ m s^{-1} at the western end of the Strait of Juan de Fuca. The high-pressure region provided a drainage air mass from the interior of British Columbia that flowed through the Straits of Georgia and Juan de Fuca and eventually into the Pacific Ocean. This air mass remained nearly homogeneous and was capped by a well-defined inversion. For the offshore low-pressure center, the lower atmosphere was stably stratified throughout the region, and

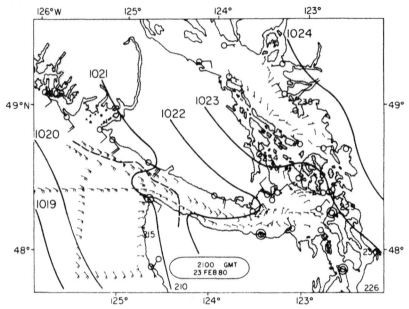

Fig. 7.23. Local wind field on February 23 and sea-level pressure analysis for 2100 GMT on February 23, 1980. The synoptic pressure field was decreasing uniformly over the region at 0.4 mb hr^{-1}. Surface wind observations are in bold print. The heavy dashed line at the entrance to the Strait of Juan de Fuca represents the transition from subsidence on the north and east side to positive vertical motion seaward of the line. [After Overland and Walter (1981).]

weak winds were observed at the eastern end of the Strait of Juan de Fuca, with strong winds at the western end. Although the features of the flow fields were complex, major characteristics of the wind fields can be accounted for by the combined effect of topography and the synoptic pressure field. Local winds were in approximate ageostrophic equilibrium between the inertia term and the imposed sea-level pressure gradient. This may be explained by a one-dimensional equation of motion:

$$\text{Terms} \qquad \frac{\partial U}{\partial t} + u\frac{\partial U}{\partial x} = fv - \frac{1}{\rho}\left(\frac{\partial P}{\partial x}\right) + K\frac{\partial^2 U}{\partial Z^2} \qquad (7.13)$$

$$\text{Scales} \qquad \frac{U}{t} \quad \frac{U^2}{L} \quad fu \quad \frac{\Delta p}{\rho L} \quad \frac{KU}{H^2}$$

$$\text{Magnitude} \quad 10^{-3} \ \ 10^{-2} \quad 10^{-3} \ \ 10^{-2} \qquad 10^{-4}$$

where typical values of scaling parameters for this type of mesoscale motion are (Overland and Walter, 1981) $u \simeq 10^1$ m s^{-1}, $L \simeq 10^4$ m, $t \simeq 10^4$ s, $H \simeq 10^3$ m, $\Delta p \simeq 10^2$ Pa, f $\simeq 10^{-4}$ s^{-1}, $\zeta \simeq 1$ kg m^{-3}, and K $\simeq 10^1$ m^2 s^{-1}.

Thus Eq. (7.13) may be simplified as

$$u \frac{\partial u}{\partial x} = -\frac{1}{\rho} \left(\frac{\partial P}{\partial x} \right)$$

or

$$\frac{\partial}{\partial x} \left(\frac{u^2}{2} \right) = -\frac{1}{\rho} \left(\frac{\partial p}{\partial x} \right) \tag{7.14a}$$

In integrated form, this gives the Bernoulli equation:

$$\frac{U^2}{2} = \frac{U_0^2}{2} - \frac{\Delta p}{\rho} \tag{7.14b}$$

where U_0 is the initial velocity and Δp is the pressure difference.

An application of Eq. (7.14) made by Overland and Walter (1981) for a constant pressure gradient of 2 mb $(100 \text{ km})^{-1}$ with $u = 0$ at $x = 20$ km is shown in Fig. 7.24 along with the observed winds and east–west pressure gradient at 0000 GMT on February 24 in the Strait of Juan de Fuca. Observed winds have a similar form to that computed for a constant pressure gradient up to just seaward of the coastline. At that point, the observed pressure continues to increase, but the wind speed remains constant, which can be interpreted as a transition toward geostrophic balance.

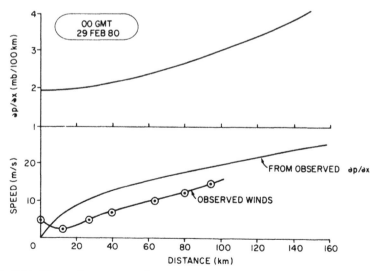

Fig. 7.24. Observed pressure gradient for the central axis of the Strait of Juan de Fuca beginning at the east end. Also shown are the observed wind speeds and the computed wind speeds from (2) for a constant pressure gradient of 2 mb $(100 \text{ km})^{-1}$. [After Overland and Walter (1981).]

Equation (7.14) can also be used to estimate the winds such as shown in Fig. 7.22. As discussed in Section 7.5.2, Reed (1980) indicates that in the vicinity of the Hood Canal Bridge on February 13, 1979, the pressure gradient was a maximum of 6 mb in 15 km or 0.4 mb km^{-1}. If we set $\Delta p = 0.4$ mb km^{-1} and note that 1 mb $= 10^2$ Pa (kg m^{-1} s^{-2}) and that Δp and Δx are in opposite sign, we have [from Eq. (7.14a)] for $x = 18$ km from the bridge

$$\frac{(U^2/2) - (U_0^2/2)}{18.5} = 0.4 \times 10^2/\rho$$

(where $\rho = 1.3$ kg m^{-3}). Thus $U = 41$ m s^{-1} when $U_0 = 22.5$ m s^{-1}. This is in excellent agreement with the observation of 40 m s^{-1} as shown in Table 7.1.

7.5.4   Diurnal Variation of the Low-Level Jet in Somalia

In Somalia, synoptic northeast monsoons generally blow from sea to land during the winter, whereas southwest winds blow from land to sea during the summer (see, e.g., Atkinson, 1971). Owing to the baroclinic field produced by the temperature difference across the coastal zone, as well as the presence in the region of mountain ranges and the Gulf of Aden, mesoscale land/sea breezes and mountain/valley winds also exist and interact to a varying degree with the synoptic flow (see, e.g., Flohn, 1965). This is particularly pronounced after the onset of the Somali low-level jet, which increases from 10–15 m s^{-1} during the day to 20–25 m s^{-1} during the night (Fig. 7.25). This low-level jet is a component of large-scale cross-equatorial flow (see, e.g., Hart et al., 1978). Krishnamurti and Wong (1979) and Bannon (1979a,b), among others, have studied the dynamics of this jet. Using pilot balloon observations, Ardanuy (1979) has investigated some of the diurnal variations of the Somali jet. The

Table 7.1

Comparison of Winds Reported by Reed (1980) in Hood Canal at 1400–1500 GMT, February 13, 1979 (cf. Fig. 7.22b), and Those Calculated from Eq. (7.14)[a]

Location	Observed (m s^{-1})	Calculated (m s^{-1})
Bridge	40	40.5
9.25 km SSW	30	33.0
18.5 km SSW	22.5	22.5[b]

[a] The initial velocity used for the calculation was 22.5 m s^{-1} (after Walter and Overland, 1982).
[b] Assumed.

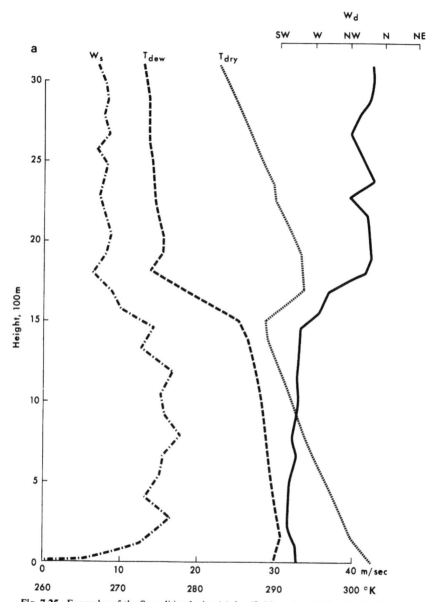

Fig. 7.25. Examples of the Somali jet during (a) day (3:16 p.m.) and (b) night (3:55 a.m. local time) on June 23, 1979, as measured at Gardo. The terms T_{dry} and T_{dew} represent dry-bulb and dew-point temperatures; W_s and W_d denote wind speed and direction, respectively.

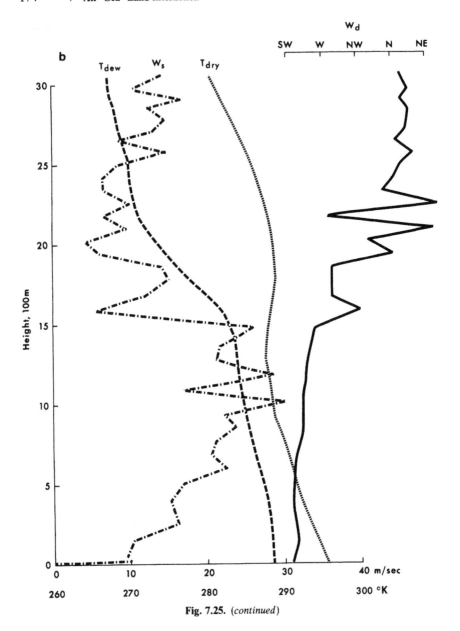

Fig. 7.25. (*continued*)

daytime evolution of the East African jet has been studied both observationally and theoretically by Rubenstein (1981).

In order to explain the diurnal variation of the low-level jet in Somalia, radiosondes were launched four times daily and tracked by theodolites at two locations in Somalia during a monsoon experiment in 1979. Analysis of upper-air data indicated that the jet consists of four distinct vertical layers: surface, adiabatic, inversion, and again adiabatic. Some examples are shown in Fig. 7.25. During the day the surface layer behaves superadiabatically; with about 10–15 m s^{-1} wind speed, it exists only between the surface and 100 m. Above this height an adiabatic layer with steady wind speeds of 15–20 m s^{-1} prevails up to 1500 m. From 1.5 to about 2.0 km a layer of subsidence inversion exists. Above 2 km adiabatic conditions again prevail. Note that the wind blows from southwest to west below the inversion layer. Wind speed shear is largest in the surface layer. Above the inversion layer, wind speed decreases to less than 10 m s^{-1}.

At night the surface layer becomes a ground-based inversion (or adiabatic) that extends to about 100 m. Within this layer, wind speed is generally less than 10 m s^{-1}. From 100 m to about 1200 m an adiabatic layer exists in which the wind speed is not as strong in either temperature or humidity as during the daytime. The adiabatic layer above this subsidence inversion remains about the same as during the day. Wind speed decreases to around 10 m s^{-1} above the inversion layer, and wind direction veers from southwest and west below the inversion layer to northwest and north above this layer.

Although reasons for diurnal variation of the jet in Somalia have been advanced by Ardanuy (1979), from Fig. 7.25 it is clear that daytime and nighttime thermodynamic structures are not very much different except in the surface layer, which extends only to about 100 m. Since the largest variation is in the wind speed, not the wind direction or temperature and humidity, differences in pressure between Mogadishu and Gardo during the day and during the night were examined. It was found that the daytime and nighttime difference in pressure gradient between these two stations is small: it amounted to only about 1 mb between 00Z and 12Z on June 13, 1979 (cf. Fig. 7.25).

Since the diurnal pressure gradient difference between these two stations is small and the Coriolis force does not change diurnally, some other mesoscale mechanism must be operating. It was found that the cooler temperature associated with the Somali current offshore, originally produced by this jet, feeds back to the atmosphere by enhancing the temperature gradient across the coastal zone. These interactions in turn affect the velocity of the land/sea breeze system along the Somali coast. This diurnal and pulsational phenomenon is further augmented by the unique setting of the Gulf of Aden and the mountain and valley winds associated with the regional geography.

The contributions of microscale forcing due to local heating and cooling as well as synoptic pressure gradient differences between day and night are small. Therefore, the combination of mesoscale land/sea breeze and mountain/valley wind systems plays a dominant role in this region. These mesoscale wind systems will retard or reduce the jet speed during the day and enhance it during the night. To evaluate the relative contribution of mesoscale winds, a sea-breeze model is utilized (cf. Section 7.1). The model was originally developed by Hauwritz (1947) and was applied successfully on the Texas coast (Hsu, 1970). For our purpose it is sufficient to note that the mean sea breeze V, due to maximum temperature difference between land (T_{land}) and the sea (T_{sea}), is

$$V = (R/L)\ln(P_0/P_1)(T_{land} - T_{sea})K(K^2 + \Omega^2)^{-1} \qquad (7.15)$$

where R is the gas constant for the air, L is the total length of the sea-breeze circulation, P_0 is the pressure near the surface, P_1 is the pressure at the upper limit of sea-breeze return flow, K is a constant that expresses the intensity of the frictional force, and Ω is the angular velocity of the earth. For more detail, see Hsu (1970). Information from satellite and radiosonding for Somalia indicated that $P_0 = 1000$ mb, $P_1 = 700$ mb, $L = 300$ km, and $K = 2 \times 10^{-5}$. Thus Eq. (7.15) becomes

$$V = 1.2(T_{land} - T_{sea}) \qquad (7.16)$$

where V is in meters per second and T is in degrees Celsius.

To explain the day–night difference in the jet shown in Fig. 7.25, temperature differences in June between land and sea were studied. It was found that during the day the value of ($T_{land} - T_{sea}$) was about 12°C and at night nearly 0. Substituting these values into Eq. (7.16), we have a 14-m s^{-1} mesoscale wind speed due to baroclinic effect operating during the day, whereas at night a barotropic field prevails across the coastal zone. Thus the jet has no "opposing" force to encounter at night and therefore the speed is higher.

It is clear that the alternation of the mesoscale baroclinic field during the day with barotropic at night is the most important mechanism to explain the diurnal variation of the jet speed as well as the interaction between these mesoscale winds and the synoptic flows.

7.5.5 Phenomenal Rainfall in Southern Nicaragua

Phenomenal rainfall occurs in some tropical coastal regions under two prominent conditions: (1) the coast is located on the windward coast with a nearby downwind mountain range and (2) a broad, shallow sea exists upwind from the coast.

The Miskito Bank, off the Atlantic coast of Nicaragua, is an excellent example (Murray *et al.*, 1982). Typical depths of only 30 m extend well over 200 km offshore. Winds across this bank are dominated by the persistence of the northeasterly trade wind system.

The rainfall rate along contiguous coastal watersheds is one of the highest in the world, exceeding 6 m (nearly 20 ft) per year along the southern end of the bank. In the Rio Escondido Basin (river number 9 in Fig. 7.26) the rainfall rate shows a high of 4.3 m year^{-1} at the coast, dropping off linearly to 3 m year^{-1} 100 km inland.

Annual rainfall at the Nicaragua–Costa Rica boundary is more than twice that at the Nicaragua–Honduras border (Fig. 7.26) (Portig, 1965, 1972). Note also on Fig. 7.26 that the coastal plain extends three to five times farther inland in the northern and central parts of the coast than at the southern end of Nicaragua.

Long-term wind data indicate that the general large-scale direction of the trade wind does not change appreciably between Puerto Cabezas (PC) and

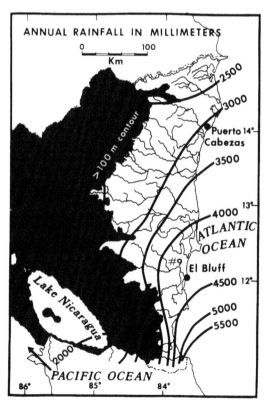

Fig. 7.26. General topography, major rivers, and annual rainfall (mm) for eastern Nicaragua (see text for explanation).

Bluefields (BF) (cf. Fig. 7.26). The wind speed, however, is stronger at PC than at BF. This orographic or "blocking" effect by the topography in the south is an important contribution to the heavy rainfall in this region.

Another important cause for this phenomenal rainfall is the atmospheric convective activity caused by intensive air–sea interaction on the broad and shallow shelf upwind from the coast. Note that under disturbed conditions in the tropics as convective activity develops, rather abrupt changes may be observed in the inversion layer and the subcloud layer underlying it. These changes persist long (up to 15 hr) after the convective systems pass (Rasmussen *et al.*, 1976). Such phenomena, with some variations, were also observed by both radiosondings and acoustic soundings near and at Bluefields in a special field experiment. For example, the intensive disturbance beginning at 0800 hr on August 22, 1976, persisted in strength for 27 hr (see Murray *et al.*, 1982).

Regional rainfall data in Central America indicate that a similar, albeit much smaller, focus of unexplained intense rainfall off Colon, Panama (e.g., Portig, 1972) is also associated with a shallow bank offshore.

7.5.6   Winter Monsoon Convection in the Vicinity of North Borneo

According to Houze *et al.* (1981), radar and satellite observations in the vicinity of northern Borneo obtained during an international winter monsoon experiment showed that the convection in that region underwent an extremely regular diurnal cycle. Over the sea to the north of Borneo, the general level of convective activity was increased during monsoon surges and during the passages of westward-propagating near-equatorial disturbances. Convective activity was decreased during monsoon lulls. The diurnal cycle was well defined, regardless of whether the general level of convective activity was enhanced or suppressed by synoptic-scale events.

The cycle of convection over the sea was especially well documented, as shown in Fig. 7.27. It was typically initiated at about midnight, when an offshore low-level wind began. Where this wind met the monsoonal northeasterly flow, usually just off the coast, convective cells formed. After midnight, the convection continued to develop, and by 0800 LST it had evolved into an organized mesoscale system with a precipitation area often continuous over a horizontal distance of 200 km. The structure of this system resembled that of squall lines and other organized mesoscale systems observed in the tropics. The precipitation was composed partially of convective cells, but a considerable portion was also stratiform with a well-defined melting layer extending across much of the system. This precipitation fell from a large mid- to upper-level cloud shield. The mesoscale systems typically began dissipating at midday, when the offshore wind reverted to an onshore wind and low-level convergence became concentrated over land.

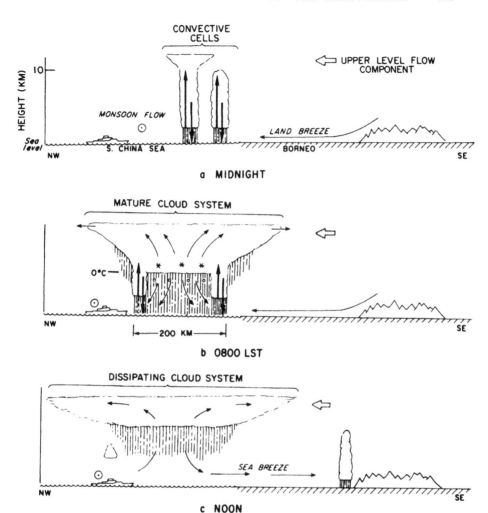

Fig. 7.27. Schematic of the development of diurnally generated mesocale precipitation feature off the coast of Borneo. Various arrows indicate airflow. Circumscribed dot indicates northeasterly monsoon flow out of page. Wide open arrow indicates the component of the typical east-southeasterly upper-level flow in the plane of the cross section. Heavy vertical arrows in (a) and (b) indicate cumulus-scale updrafts and downdrafts. Thin arrows in (b) and (c) show a mesoscale updraft developing in a mid- to upper-level stratiform cloud with a mesoscale downdraft in the rain below the middle-level base of the stratiform cloud. Asterisks and small circles indicate ice above the 0°C level melting to form raindrops just below this level. [After Houze et al. (1981).]

Chapter 8 | Engineering Meteorology

Many engineering projects require meteorological inputs. This is particularly true in the coastal zone. For example, wind and wave information are vital in the design of coastal structures. Refractivity, resulting from humidity anomaly, to a large extent affects the performance of microwave communication systems, and characteristics of the internal boundary layer across the shoreline are the controlling factor in the dispersion and transport of air pollutants.

This chapter is devoted to topics that deal with meteorological characteristics in the coastal zone. This knowledge is needed in order to improve the design of engineering systems.

8.1 Estimating Wind Speeds for Offshore Applications

It is well known that differences exist between onshore and offshore winds, particularly wind speeds. Examples are shown in Fig. 8.1, which indicates that, contrary to common practice, even stations such as those located on a small flat, open area (e.g., Key West, Florida, and Cape Hatteras, North Carolina) cannot represent monthly mean offshore conditions. Furthermore, at stations located in an estuarine environment such as Homer, in Cook Inlet, Alaska, the mean wind speed on the average is less than half that offshore. Figure 8.1 also shows that the largest difference between onshore and offshore conditions usually occurs in winter, when offshore storms are more frequent. During the warmer part of the year higher speeds may be recorded, at times, for diurnal onshore (sea breeze) winds than for offshore winds. As expected the wind speed decreases inland.

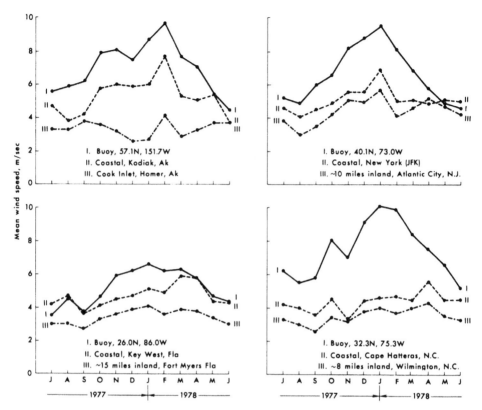

Fig. 8.1. Examples of differences in monthly wind speed (m s^{-1}) at stations ranging from offshore to coastal to inland (from July 1977 through June 1978) (Hsu, 1981b). Copyright © by D. Reidel Publishing Co. Reprinted by permission.

Many studies related to coastal marine sciences and engineering require wind data from offshore regions. Yet *in situ* measurements over water are often lacking. Therefore, engineers as well as scientists traditionally rely on wind measurements over land, preferably near coasts. The correction of land-based wind data for offshore applications is of major concern to all of us working in the coastal zone. A formula that linearly relates the difference in wind speed between onshore and offshore regions, as tested successfully in the Great Lakes region, has been revised and extended to other parts of the world (Hsu, 1986b). This formula is further substantiated theoretically by using an approximation of the equations of motion. Contribution of air–sea temperature difference to wind speed and direction, as well as the meteorological conditions under which this formula may be applied, are also evaluated.

In order to facilitate corrections to inland station data before applying them to offshore regions, a theoretical formula is developed and proposed, as follows (Hsu, 1981b):

In the atmospheric planetary boundary layer (PBL) one may assume that (1) horizontal mean wind shears are small compared to vertical mean wind shears, and (2) a balance exists among acceleration, Coriolis, pressure gradient, and eddy-viscosity forces at every level.

Under these conditions the horizontal equations of motion may be written (see, e.g., Hess, 1959, p. 280; or Holton, 1979, p. 105):

$$\frac{du}{dt} = fv - \frac{1}{\rho}\left(\frac{\partial P}{\partial x}\right) + \frac{1}{\rho}\left(\frac{\partial \tau_{zx}}{\partial Z}\right) \tag{8.1}$$

$$\frac{dv}{dt} = -fu - \frac{1}{\rho}\left(\frac{\partial P}{\partial y}\right) + \frac{1}{\rho}\left(\frac{\partial \tau_{zy}}{\partial Z}\right) \tag{8.2}$$

where u and v are the wind components along the x and y coordinates, respectively, Z is the vertical axis, f is the Coriolis parameter, ρ is density, P is pressure, and τ_{zx} and τ_{zy} are the eddy stresses.

Without loss of generality we may further assume for the wind across the coastal zone that

1. The acceleration terms du/dt and dv/dt are small compared to other terms in Eqs. (8.1) and (8.2).
2. The direction follows the x axis near the surface (i.e., $v = 0$ and therefore $\tau_{zy} = 0$).
3. There is a geostrophic wind on top of the PBL with its components U_g and V_g, where

$$U_g \equiv \frac{1}{\rho f}\left(\frac{\partial P}{\partial y}\right) \qquad V_g \equiv \frac{1}{\rho f}\left(\frac{\partial P}{\partial x}\right)$$

Note that owing to the effect of the Ekman spiral within the PBL, the direction of the surface wind is not the same as that of the geostrophic wind. In other words, if u is along the x axis near the surface, where $v = 0$, as assumed previously, neither U_g nor V_g is zero in magnitude, but they may be considered as variables that do not change appreciably on top of the PBL across the coastal zone.

Under these conditions we have

$$\partial \tau_{zx}/\partial Z = \rho f V_g \tag{8.3}$$

Following Mahrt and Lenschow (1976), we now integrate Eq. (8.3) with respect to Z from a reference level, say $Z = 10$ m above the surface, to the top of the PBL with height H. Note that the eddy stress at the bottom of the surface layer is parameterized with a drag law, employing a drag coefficient,

$C_D \equiv (U_*/U_z)^2$ (where U_* is the friction velocity and U_z is the wind speed at height Z; see, e.g., Mahrt and Lenschow, 1976). Equation (8.3) then becomes

$$\int_{\tau_{surface}}^{\tau_{top}} d\tau = \int_{10\,m}^{H} \rho f V_g \, dZ = \rho f V_g \int_{10\,m}^{H} dZ$$

or

$$\frac{\tau_{top} - \tau_{surface}}{H - 10 \quad m} = \rho f V_g \tag{8.4}$$

Since $\tau_{top} = 0$, $\tau_{surface} = \rho C_D U^2$, and $H \gg 10$ m (since the height τ of the PBL normally is in hundreds of meters), we have

$$\frac{C_D U^2}{H} = -f V_g \tag{8.5}$$

Note that the right-hand side of Eqs. (8.3), (8.4), and (8.5) consists of the density, the Coriolis parameter, and the geostrophic component. These terms may be considered constant across the coastal zone. We then have

$$\frac{(C_D U^2/H)_{land}}{(C_D U^2/H)_{sea}} = 1 \tag{8.6}$$

or

$$\frac{U_{sea}}{U_{land}} = \left[\frac{H_{sea} C_{D,\,land}}{H_{land} C_{D,\,sea}}\right]^{1/2} \tag{8.7}$$

where U, H, and C_D are the wind speed, height of the planetary boundary layer (PBL), and drag coefficient, respectively. Subscripts sea and land represent offshore and onshore conditions, respectively. Equation (8.7) should be applicable whether the wind blows from land to sea or vice versa, assuming that the geostrophic wind above the PBL across the coastal zone does not change appreciably.

For a given climatological regime, values of parameters on the right-hand side of Eq. (8.7) are known. For example, using data from Holzworth (1972) and SethuRaman and Raynor (1980) for the New York–Massachusetts region (see Hsu, 1981b) and Eq. (8.7), $U_{sea}/U_{land} = 1.7$, where $H_{sea} = 620$ m, $H_{land} = 1014$ m, $C_{D,\,land} = 0.0083$, and $C_{D,\,sea} = 0.0017$ (at 8 m above the land and sea surfaces). Therefore, from Eq. (8.7) and this example, U_{sea} is linearly related to U_{land}. From a statistical point of view, we then have

$$U_{sea} = A + B U_{land} \tag{8.8}$$

where values of A and B are to be determined empirically.

Note that since a low-level jet may prevail over a coastal region near the surface, particularly in the offshore region (see, e.g., Hsu, 1979a), U_{sea} may not be equal to zero when U_{land} is zero. In other words, when onshore conditions are calm, it is not necessary that winds offshore also be calm. This is because strong pressure gradient and baroclinic effect exist across the coastal zone, so that during the day there might be a sea breeze and at night stronger winds such as a land breeze or low-level jet might prevail offshore (Hsu, 1970, 1979a). Thus parameter A in Eq. (8.8) is also a necessary meteorological requirement.

Recently, Liu *et al.* (1984) refined equations originally developed by Schwab (1978) based on graphs given by Resio and Vincent (1977) for the Great Lakes region when the wind was blowing from land to water:

$$\frac{U_{sea}}{U_{land}} = \left(1.2 + \frac{1.85}{U_{land}} \right) \phi(\Delta T) \tag{8.9}$$

$$\phi(\Delta T) = 1 - \frac{\Delta T}{|\Delta T|} \left(\frac{|\Delta T|}{1920} \right)^{1/3} \tag{8.10}$$

$$\Delta \theta = (12.5 - 1.5 \Delta T) - (0.38 - 0.03 \Delta T)U_{sea} \tag{8.11}$$

where ΔT is the air–water temperature difference (°C) and $\Delta \theta$ is the clockwise angle between over-land and over-lake winds (degrees).

Equations (8.9), (8.10), and (8.11) have been successfully applied to modeling storm-surge and current fluctuations in the Great Lakes (Schwab, 1978, 1983)

Equations (8.10) and (8.11) are related to the correction of the ratio of U_{sea}/U_{land} by air–sea temperature difference and the difference in wind direction between land and sea, respectively. If the corrections predicted by Eqs. (8.10) and (8.11) can be shown to be smaller than the reported differences resulting from instrumental noise and recorder errors, one may neglect these differences. This was done, and is demonstrated as follows.

Equations (8.10) and (8.11) are plotted in Figs. 8.2 and 8.3, respectively. It can be seen for $|\Delta T| = 5°C$, the correction to U_{sea}/U_{land} is less than 15% and the value of $\Delta \theta$ is less than 18°. Also, the contribution of U_{sea} to $\Delta \theta$ is small. Note that even when T_{sea} is 20°C warmer than the air, $\Delta \theta \simeq 38°$.

On the other hand, it has been shown by Haltiner and Martin (1957, p. 235) that the difference in values of the surface cross-isobar angle between onshore and offshore airflow can be 20°. Furthermore, according to Mazzarella (1985, p. 291), the field accuracy for wind direction for a common wind vane is approximately 8°, and for gust direction 15°. The recorded resolution for the wind vane is 10°. Therefore, the composite contribution by frictional effects, instrument error, and recorder inaccuracy may reach 45° in wind direction. In other words, for the same geostrophic wind across the coastal zone, the difference in wind direction between onshore and offshore near the surface may be as large as 45°. For this reason, and because most of the time $|\Delta T|$ is much less than 20°C, corrections of $\Delta \theta$ by U_{sea} and ΔT are not very important operationally, as shown in Fig. 8.3.

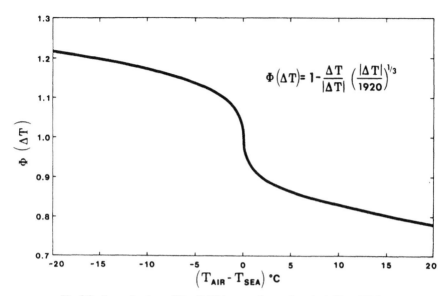

$$\Phi(\Delta T) = 1 - \frac{\Delta T}{|\Delta T|} \left(\frac{|\Delta T|}{1920} \right)^{1/3}$$

Fig. 8.2. An evaluation of Eq. (8.10) (see text for explanation) (Hsu, 1986b).

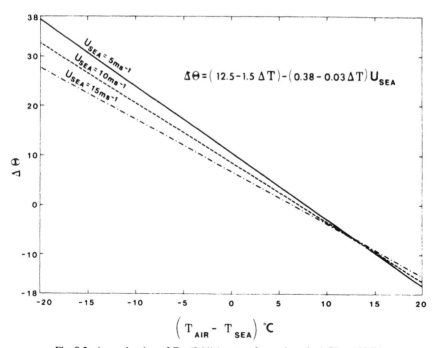

$$\Delta\Theta = (12.5 - 1.5\,\Delta T) - (0.38 - 0.03\,\Delta T)\,U_{SEA}$$

Fig. 8.3. An evaluation of Eq. (8.11) (see text for explanation) (Hsu, 1986b).

From the above evaluation a criterion may be set: if the difference in direction between onshore near-surface winds and offshore near-surface winds is within 45°, Eq. (8.7) may be employed, whether the wind blows from land to sea or vice versa. From a meteorological point of view, then, this criterion should be applicable under weather systems such as land and sea breezes (e.g., Hsu, 1970), hurricanes near landfall (Powell, 1982), and other larger (or synoptic-scale) phenomena such as cyclones (lows), monsoons, and anti-cyclones (highs), but not during the passage of atmospheric fronts and squall lines across the coastal zone. This criterion also implies that climatological wind data onshore (say, monthly averages) would be used for offshore estimates, since the transient weather systems, such as squall lines, frontal passages, and local thunderstorms, usually do not last more than a day or two. If we assume that ΔT is small in Eq. (8.9) (Fig. 8.2), say less than $\pm 5°C$, as normally is the case, and since the aggregate wind estimation error cannot be less than 10% at airports, where most official weather service stations are located (see Wieringa, 1980), we have (by setting $\phi(\Delta T) \simeq 1$)

$$\frac{U_{sea}}{U_{land}} = 1.2 + \frac{1.85}{U_{land}}$$

or

$$U_{sea} = 1.85 + 1.2 U_{land} \tag{8.12}$$

This further supports the general form of Eq. (8.8). On the other hand, one may state that the semiempirical formula shown in Eq. (8.12) is supported by our theoretical considerations, shown in Eq. (8.7).

Based on many pairs of data sets, from environments ranging from the tropics to the Gulf of Alaska, Eq. (8.8) is verified as follows.

Many pairs of onshore and offshore measurements under land and sea breezes, monsoon, and synoptic-scale weather systems have become available recently (Hsu, 1981b). In addition to those listed in Hsu (1981b), one more pair, obtained under hurricane landfall conditions, is included in Hsu (1986b). Ratios of U_{sea}/U_{land} were analyzed as a function of U_{land}. Note that in Hsu (1981b) wind speeds were below 18 m s^{-1}. The data set provided in Powell (1982) included hurricane-force wind measurements obtained during Hurricane Frederic, which struck the Alabama coast in 1979 (cf. Chapter 5).

Pairs of U_{sea} and U_{land} were linearly regressed as guided by Eq. (8.8). The result is plotted in Fig. 8.4, and the linear regression equation is

$$U_{sea} = 1.62 + 1.17 U_{land} \tag{8.13}$$

with a very high value of coefficient of determination, $r^2 = 0.99$.

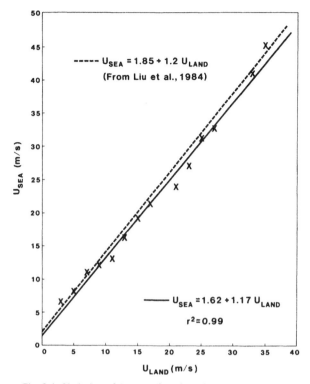

Fig. 8.4. Variation of U_{sea} as a function of U_{land} (Hsu, 1986b).

For comparison, Eq. (8.12) is also plotted in Fig. 8.4. It is shown that the difference between Eqs. (8.12) and (8.13) is not significant; for example, at the highest wind speed observed, $U_{land} = 35 \text{ m s}^{-1}$, the difference between the values of U_{sea} is less than 1.5 m s^{-1}. Since Eq. (8.13) did not take air–sea temperature differences into consideration, these differences are surprisingly negligible. Equation (8.13) is thus recommended for operational use. If $|\Delta T|$ is large, say, $5°C$, Eqs. (8.9) and (8.10) (or Fig. 8.2) may be employed. Equation (8.13) should also be useful for climatological applications.

However, for the more accurate forecasting required by marine meteorologists at the National Weather Service, Eq. (8.7) should be used. The value of H_{sea} may be estimated from H_{land} (Hsu, 1979b), which is routinely available from radiosonde measurements. To obtain the value of $C_{D, land}$, one may use 10×10^{-3}, as obtained by Garratt (1977). For values of $C_{D, sea}$, other studies may be consulted, such as Large and Pond (1981) (cf. Chapter 6).

8.2   Estimates of Atmospheric Dispersion

Atmospheric dispersion in the coastal zone requires some modification in techniques obtained from inland experiments. Two important considerations are (1) over water stability is much less pronounced than that over land and (2) the internal boundary layer develops over the shoreline because the airflow encounters changes in both roughness and temperature structures.

This section provides some common techniques for estimating the atmospheric dispersion on land, over water, and at shorelines, inasmuch as they are different in these three coastal environments.

8.2.1   An Operational Formula for Dispersion Estimates

Many formulas and models have been devised for dispersion estimates, but the most common is based on Gaussian or normal distribution (see, e.g., Turner, 1969). Before this formula is applied, some derivations are needed. The following discussion mainly follows Hoel (1965).

The normal or Gaussian distribution is defined as

$$f(x) = ce^{-(1/2)(x/b)^2} = c\exp\left[-\left(\frac{1}{2}\right)\left(\frac{x}{b}\right)^2\right] \tag{8.14}$$

where a, b, and c are parameters that make $f(x)$ a frequency function.

A representation of Eq. (8.14) is shown in Fig. 8.5. Note that if $f(x)$ is the frequency function of the random variable x, the moment-generating function of $g(x)$ is given by

$$M_{g(x)}(\theta) = \int_{-\infty}^{\infty} \exp \theta g(x) f(x)\, dx \tag{8.15}$$

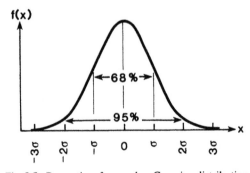

Fig. 8.5. Properties of normal or Gaussian distribution.

From Eq. (8.15), setting $g(x) = x$, we have

$$M_x(\theta) = c \int_{-\infty}^{\infty} \exp \theta x \exp\left[-\left(\frac{1}{2}\right)\left(\frac{x}{b}\right)^2 \right] dx$$

Let $z = x/b$; then $dx = b\,dz$ and

$$M_x(\theta) = bc \int_{-\infty}^{\infty} \exp(\theta bz - z^2/2)\,dz$$

because

$$\theta bz - z^2/2 = -\tfrac{1}{2}(z - \theta b)^2 + \tfrac{1}{2}\theta^2 b^2$$

Then

$$M_x(\theta) = bc \exp \tfrac{1}{2}\theta^2 b^2 \int_{-\infty}^{\infty} \exp[-\tfrac{1}{2}(z - \theta b)^2]\,dz$$

If $t = z - \theta b$, then $dz = dt$ and

$$M_x(\theta) = bc \exp[\tfrac{1}{2}\theta^2 b^2] \int_{-\infty}^{\infty} \exp(-t^2/2)\,dt$$

Now let

$$I = \int_0^{\infty} \exp(-t^2/2)\,dt$$

Then

$$I^2 = \int_0^{\infty} \exp(-x^2/2)\,dx \int_0^{\infty} \exp(-y^2/2)\,dy$$

$$= \int_0^{\infty} \int_0^{\infty} \exp[(x^2 + y^2)/2]\,dx\,dy$$

In polar coordinates this double integral assumes the form

$$I^2 = \int_0^{\pi/2} \int_0^{\infty} \exp(-r^2/2)\,r\,dr\,d\theta$$

$$= \int_0^{\pi/2} -\exp(-r^2/2)\big|_0^{\infty}\,d\theta$$

$$= \int_0^{\pi/2} d\theta = \pi/2$$

So

$$\int_{-\infty}^{\infty} \exp(-t^2/2)\, dt = (2\pi)^{1/2}$$

and

$$M_x(\theta) = (2\pi)^{1/2} bc \exp(\tfrac{1}{2}\theta^2 b^2) \tag{8.16}$$

Note that for any moment-generating function $M(0) = 1$; hence from Eq. (8.16) it follows that $bc(2\pi)^{1/2} = 1$ and that

$$M_x(\theta) = \exp(\tfrac{1}{2}\theta^2 b^2)$$

If this exponential is expanded in a power series, then

$$M_x(\theta) = 1 + b^2 \frac{\theta^2}{2} + \cdots$$

Note that the coefficient of $\theta^2/2!$ is the second moment of x about its mean; therefore, $b^2 = \sigma^2$, or $b = \sigma$, where σ is the standard deviation. Since $(2\pi)^{1/2} bc = 1$, $c = 1/\sigma\sqrt{2\pi}$; therefore, Eq. (8.14) can be written in the form

$$f(x) = \frac{1}{\sigma(2\pi)^{1/2}} \exp\left[-\frac{1}{2}\left(\frac{x}{\sigma}\right)^2 \right] \tag{8.17}$$

According to Hoel (1965), if two random variables y and z are normally distributed but in addition are independently distributed, then their joint frequency function is the product of the two marginal frequency functions, that is,

$$f(y,z) = \frac{\exp[-(1/2)(y/\sigma_y)^2]}{(2\pi)^{1/2}\sigma_y} \frac{\exp[-(1/2)(z/\sigma_z)^2]}{(2\pi)^{1/2}\sigma_z}$$

or

$$f(y,z) = (2\pi\sigma_x\sigma_y)^{-1} \exp[-(1/2)(y/\sigma_y)^2] \times \exp[-(1/2)(z/\sigma_z)^2] \tag{8.18}$$

In order to preserve dimensions and to take into account the fact that the mean wind stretches the plume forward with resulting dilution (Munn, 1966), the binormal probability distribution $f(y,z)$ may be replaced by $u\chi/Q$ where u is the mean wind in the x direction, Q is the emission rate, and χ is the concentration downwind. A coordinate system for Eq. (8.18) is shown in Fig. 8.6. According to Turner (1969), in the system considered here the origin is at ground level at or beneath the point of emission, and the x axis extends horizontally in the direction of the mean wind. The y axis is in the horizontal plane perpendicular to the x axis, and the z axis extends vertically. The plume

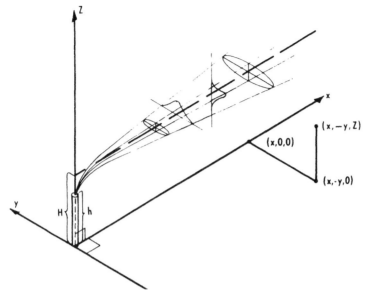

Fig. 8.6. Coordinate system showing Gaussian distributions in the horizontal and vertical. [After Turner (1969).]

travels along or parallel to the x axis. With this coordinate system, Eq. (8.18) becomes

$$\frac{u\chi}{Q} = \frac{1}{2\pi\sigma_y\sigma_z} \exp\left[-\frac{1}{2}\left(\frac{y}{\sigma_y}\right)^2\right] \exp\left[-\frac{1}{2}\left(\frac{z}{\sigma_z}\right)^2\right] \qquad (8.19)$$

Equation (8.19) is not yet complete because vertical diffusion is bounded by the surface of the earth. It is assumed that the ground acts as a perfect reflector and that there is a mirror image source at $z = -H$ as well as one at $z = +H$, as illustrated by Munn (1966, Fig. 39). For diffusion estimates, then, the concentration χ of gas or aerosols (particles less than about 20 μm in diameter) at x, y, and z from a continuous source with an effective emission height H is given by Turner (1969):

$$\chi(x, y, z; H) = \frac{Q}{2\pi\sigma_y\sigma_z u} \exp\left[-\frac{1}{2}\left(\frac{y}{\sigma_y}\right)^2\right]$$
$$\times \left\{\exp\left[-\frac{1}{2}\left(\frac{z - H}{\sigma_z}\right)^2\right] + \exp\left[-\frac{1}{2}\left(\frac{z + H}{\sigma_z}\right)^2\right]\right\} \qquad (8.20)$$

for χ(g m^{-3} or, for radioactivity, Ci m^{-3}), Q (g s^{-1} or Ci s^{-1}), U (m s^{-1}), σ_y,

σ_z, h, x, y, and z (m), and where H is the height of the plume centerline (Fig. 8.6) when it becomes essentially level, and is the sum of the physical stack height, h and the plume rise ΔH. The following assumptions are made: the plume spread has a Gaussian distribution (see Fig. 8.5) in both the horizontal and vertical planes, with standard deviations of plume concentration distribution in the horizontal and vertical of σ_y and σ_z, respectively; the mean wind speed affecting the plume is u; the uniform emission rate of pollutants is Q; and total reflection of the plume takes place at the earth's surface—that is, there is no deposition or reaction at the surface.

For concentrations calculated at ground level (i.e., $z = 0$), the equation simplifies to

$$\chi(z, y, 0; H) = \frac{Q}{\pi \sigma_y \sigma_z u} \exp\left[-\frac{1}{2}\left(\frac{y}{\sigma_y}\right)^2 \right] \exp\left[-\frac{1}{2}\left(\frac{H}{\sigma_z}\right)^2 \right] \qquad (8.21)$$

Where the concentration is to be calculated along the centerline of the plume ($y = 0$), further simplification results in

$$\chi(x, 0, 0; H) = \frac{Q}{\pi \sigma_y \sigma_z u} \exp\left[-\frac{1}{2}\left(\frac{H}{\sigma_z}\right)^2 \right] \qquad (8.22)$$

For a ground-level source with no effective plume rise ($H = 0$),

$$\chi(x, 0, 0; 0) = \frac{Q}{\pi \sigma_y \sigma_z u} \qquad (8.23)$$

If a stable layer with height L exists above an unstable layer, the concentration for any height between the ground and L can be calculated from

$$\chi(x, y, z; H) = \frac{Q}{\sqrt{2\pi}\, \sigma_y L u} \exp\left[-\frac{1}{2}\left(\frac{y}{\sigma_y}\right)^2 \right] \qquad (8.24)$$

for any z from 0 to L, for $x > 2x_L$; X_L is where $\sigma_z = 0.47L$.

8.2.2 Dispersion Estimates on Land

In Eq. (8.20) the standard deviations of the cross-wind concentration distributions in the lateral direction (σ_y) and in the vertical direction (σ_z) are functions of downwind distance x and atmospheric stability. For dispersion estimates on land, Tables 8.1 and 8.2 can be employed.

Note that the formulas in Table 8.2 are restricted to downwind distances less than approximately 10 km and applicable to simple point sources over homogeneous, flat terrain. Air-quality modeling over long distances may be found elsewhere (see, e.g., Venkatram, 1985).

Table 8.1

Pasquill's Stability Categories[a] Applicable on Land

Surface wind speed at 10 m (m s^{-1})	Day			Night	
	Incoming Solar Radiation			Thinly overcast or $\geq 4/8$ low cloud	$\leq 3/8$ cloud
	Strong	Moderate	Slight		
<2	A	A–B	B		
2–3	A–B	B	C	E	F
3–5	B	B–C	C	D	E
5–6	C	C–D	D	D	D
>6	C	D	D	D	D

[a] The neutral class, D, should be assumed for overcast conditions during day or night. "Strong" incoming solar radiation corresponds to a solar altitude greater than 60° with clear skies; "slight" insolation corresponds to a solar altitude from 15° to 35° with clear skies. Table 170, Solar Altitude and Azimuth, in the Smithsonian Meteorological Tables (List, 1984), can be used in determining the solar altitude.

8.2.3 Dispersion Estimates over Water

Atmospheric diffusion in the offshore region has received more attention recently owing to the increased numbers of oil and natural gas platforms as well as the incineration of chemical wastes by ships. In order to convert the Pasquill classes for use at sea and transfer the experience gained over land,

Table 8.2

Formulas for $\sigma_y(x)$ and $\sigma_z(x)$ ($10^2 < x < 10^4$ m)[a]

Pasquill type	$\sigma_y(m)$	$\sigma_z(m)$
	Open-country conditions	
A	$0.22x(1 + 0.0001x)^{-1/2}$	$0.20x$
B	$0.16x(1 + 0.0001x)^{-1/2}$	$0.12x$
C	$0.11x(1 + 0.0001x)^{-1/2}$	$0.08x(1 + 0.0002x)^{-1/2}$
D	$0.08x(1 + 0.0001x)^{-1/2}$	$0.06x(1 + 0.0015x)^{-1/2}$
E	$0.06x(1 + 0.0001x)^{-1/2}$	$0.03x(1 + 0.0003x)^{-1}$
F	$0.04x(1 + 0.0001x)^{-1/2}$	$0.016x(1 + 0.0003x)^{-1}$
	Urban conditions	
A–B	$0.32x(1 + 0.0004x)^{-1/2}$	$0.24x(1 + 0.001x)^{1/2}$
C	$0.22x(1 + 0.0004x)^{-1/2}$	$0.20x$
D	$0.16x(1 + 0.0004x)^{-1/2}$	$0.14x(1 + 0.0003x)^{-1/2}$
E–F	$0.11x(1 + 0.0004x)^{-1/2}$	$0.08x(1 + 0.00015x)^{-1/2}$

[a] From Briggs (1973).

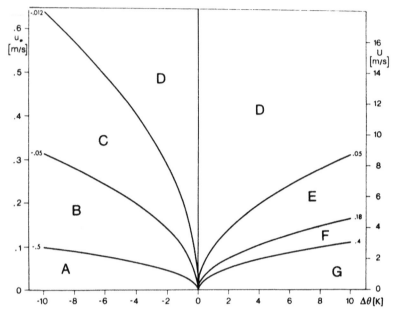

Fig. 8.7. Pasquill's stability categories transposed for use at sea: $\Delta\theta$ is the air–sea temperature difference (virtual potential temperature, if available), U is mean wind speed at about 10 m height. Curved lines are constant $1/L$ as indicated. [After Hasse and Weber (1985). Copyright © by D. Reidel Publishing Co. Reprinted by permission.]

Hasse and Weber (1985) have transposed stability categories as shown in Fig. 8.7. Their assumption is that the turbulence intensity and hence diffusion will be the same over land and over sea if the boundary conditions expressed by sensible heat flux and shear stress are equal, and with reasonable accuracy synoptic parameters, mean wind speed, and air–sea temperature difference can be used at sea to specify stability categories.

With the aid of Fig. 8.7 diffusion equations provided in the preceding section may be applied for offshore regions. Parameters σ_y and σ_z for open-country conditions as listed in Table 8.2 may be used.

8.2.4   Dispersion Estimates at Shorelines

When air flows from offshore to onshore, it is modified by changes in roughness, which produce a mechanical internal boundary layer (see, e.g., SethuRaman and Raynor, 1980), and/or by temperature contrasts, which produce a thermal or convective internal boundary layer (CIBL), as shown in Fig. 8.8 (see, e.g., Lyons, 1975). There are many studies related to the

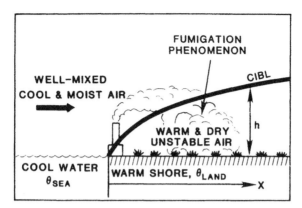

Fig. 8.8. Schematic diagram of the thermally modified boundary layer, or convective internal boundary layer (CIBL), over a warm shore caused by advection of cool air. (For explanation of symbols, see text.)

investigation of these internal boundary layers, particularly with respect to air-pollution problems in coastal environments. Lists of these studies can be found in Ogawa and Ohara (1985) and Stunder and SethuRaman (1985).

There are very few measurements relating to estimation of the height of the CIBL, particularly near shorelines. It is the purpose of this section to adapt for the CIBL case a formula that has been tested successfully for over-water conditions (Hsu, 1983a, 1984a). Knowledge of the CIBL height is also very important from the microwave routing (path) point of view because anomalous radio propagation exists as a result of atmospheric refractivity changes across the coastal zone, mainly owing to spatial humidity gradients (see, e.g., Hsu, 1983a).

From first principles, Venkatram (1977) has derived a theoretical formula for the height (h) of the CIBL:

$$h = \frac{U_*}{U_m}\left[\frac{2(\theta_{\text{land}} - \theta_{\text{sea}})X}{\gamma(1 - 2F)}\right]^{1/2} \tag{8.25}$$

where U_* and U_m are the friction velocity and mean wind speeds, respectively, inside the CIBL, γ is the lapse rate above the boundary layer or upwind condition, F is an entrainment coefficient, which ranges from 0 to 0.22, θ_{land} and θ_{sea} are the potential air temperatures over land and water, respectively, and X is the distance or fetch downwind from the shoreline. Note that in this study h is defined as the distance between the ground and the base of the CIBL (cf. Fig. 8.8).

Equation (8.25) may be simplified by employing the drag coefficient C_D, which is defined as $(U_*/U_m)^2$. Thus

$$h = \left[\frac{2C_D(\theta_{land} - \theta_{sea})X}{\gamma(1 - 2F)} \right]^{1/2} \tag{8.26}$$

Equation (8.26) will be shown to be verified by observations.

Note that the dependency of h on $X^{1/2}$ has been predicted by dimensional analysis (see, e.g., Raynor et al., 1975) and by the thermodynamic approaches of Tennekes (1973) (Mizuno, 1982; Steyn and Oke, 1982). Note also that, although there are other formulas to compute the height of the CIBL (see, e.g., Stunder and SethuRaman, 1985), Eq. (8.25) is theoretically sound because it is based on the two-dimensional mixed-layer energy equation.

Three pertinent experiments relating to estimation of the height of the CIBL are synthesized for this study: Druilhet et al. (1982), Smedman and Högström (1983), and Ogawa and Ohara (1985). Although the two data sets provided in Stunder and SethuRaman (1985) did not have air temperatures over land and water, their C_D value, based on U_* and U_m measurements, is employed here.

Equation (8.26) can be simplified as

$$h = AX^{1/2} \tag{8.27}$$

where

$$A = \left[\frac{2C_D(\theta_{land} - \theta_{sea})}{\gamma(1 - 2F)} \right]^{1/2} \tag{8.28}$$

Note that Eq. (8.27) has also been preferred by several coastal dispersion modelers (e.g., Van Dop et al., 1979; Mistra, 1980). Throughout the following calculations, values of $C_D(= 12 \times 10^{-3}$ from $U_* = 0.5$ m s^{-1} and $U = 4.5$ m s^{-1}) are based on Stunder and SethuRaman (1985) and $F = 0.2$ (Driedonks, 1982). At a site along the French Mediterranean coast, Druilhet et al. (1982, Fig. 4) have shown that average $(\theta_{land} - \theta_{sea})$ is about 2°C and $\gamma \simeq 1$°C/100 m. Substituting these values into Eq. (8.28), we have $A = 2.83$. For the flat coastal site at Näsudden, on the island of Gotland, Sweden, Smedman and Högström (1983, Fig. 2) provided the following parameters: For run N1, $U_* = 0.41$ m s^{-1}, $U_{10 m} = 3.7$ m s^{-1} (thus $C_D = 12 \times 10^{-3}$), $\theta_{land} - \theta_{sea} = 0.8$°C (between the lowest observation point and the top of the CIBL), and $\gamma = 0.8$°C/20 m. Thus $A = 0.90$. Similarly, for run N2, $U_* = 0.53$ m s^{-1}, $U_{10 m} = 5.3$ m s^{-1} (thus $C_D = 10 \times 10^{-3}$), $(\theta_{land} - \theta_{sea}) = 1$°C, and $\gamma = 0.9$°C/42 m, we have $A = 1.25$. The average A for these two runs is 1.08. Note that the value of $C_D (= 12 \times 10^{-3})$ based on Stunder and SethuRaman (1985), as stated above, is consistent not only with the Swedish data set but also with $C_D = 10 \times 10^{-3}$ from Garratt (1977).

A study by Ogawa and Ohara (1985, Fig. 3) on the coast of Japan yields $\gamma = 1°C/100$ m, as obtained at $X = 0$ m by Kitoon observations. The average value of $(\theta_{land} - \theta_{sea})$ between the inside and the outside of the CIBL at about 3 m above the surface is approximately $(25.5 - 24.5)$, or 1°C, yielding $A = 2.00$. The averaged value of A for all three coastal sites is 1.97.

In practice, however, particularly for engineering applications, parameters needed to compute the coefficient A as formulated in Eq. (8.28) are rarely measured. Therefore, an analysis of existing data sets in the following manner is warranted.

Since the variation of X ranges from 30 m to 6000 m, a log–log plot is convenient. From Eq. (8.27) we have

$$\ln h = \ln A + \tfrac{1}{2}\ln X$$

Thus

$$A = \exp[\ln h - \tfrac{1}{2}\ln X] \tag{8.29}$$

An evaluation of this parameter A from available observations is made in Tables 8.3 and 8.4. The average for all three coasts is

$$h = (1.91 \pm 0.38)X^{1/2} \tag{8.30}$$

where h and X are in meters. This equation is shown in Fig. 8.9.

In order to further compare the observation with Eq. (8.30), the 95% confidence limit is plotted also, that is, $h = (1.91 \pm 2 \times 0.38)X^{1/2}$, or $h = 2.67X^{1/2}$ and $h = 1.15X^{1/2}$, respectively.

It can be seen that all measurements are within this limit. Since the Swedish measurements (Smedman and Högström, 1983) were around 1500 m from the

Table 8.3

An Evaluation of Parameter A Based on Eq. (8.29)[a,b]

X (m)	30[c]	90[c]	160[c]	1000[d]	1500[e]	2000[d]	3000[d]	4000[d]	5000[d]	6000[d]
h (m)	8	11	24	79	76	105	123	138	151	162
A	1.461	1.160	1.897	2.498	1.962	2.348	2.246	2.182	2.135	2.091

[a] From Hsu (1986c).

[b] X, Fetch or distance downwind from the shoreline; h, height of the convective internal boundary layer; A is obtained from Eq. (8.29).

[c] Observations from a Japanese coast based on Fig. 3 in Ogawa and Ohara (1985).

[d] Measurements from a French coast based on Fig. 8b in Druilhet *et al.* (1982) and reduced from their equation $h \approx 5X^{0.4}$.

[e] Observations from a Swedish coast based on the parameter Z_i in Table 1 in Smedman and Högström (1983). Note that the mean and standard deviation of Z_i for all nine runs are 76 ± 13. Note also that it is felt by this author that Z_i is more representative in our case because heat flux is 0 at that level rather than the values obtained by slab modeling.

Table 8.4

Comparison of Parameter A from Three
Different Coasts[a]

Coast	$A \pm$ SD[b]
Japanese coast	1.506 ± 0.371
Swedish coast	1.962
French coast	2.250 ± 0.151
Average of all three coasts (equally weighted)	1.906 ± 0.375

[a] From Hsu (1986c). Also see Table 8.3.
[b] Standard deviation.

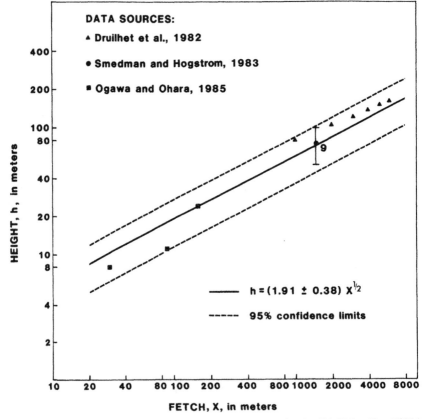

Fig. 8.9. Variation of h on X based on Eq. (8.27) (see text for details). [After Hsu (1986c). Copyright © by D. Reidel Publishing Co. Reprinted by permission.]

shoreline and there were nine runs, the mean and two standard deviations (i.e., 95% confidence limit) are plotted in Fig. 8.9 also. These limits are within the 95% bounds.

The above analyses indicate that Eq. (8.30) is a good approximation for practical applications. Note that the constant A as shown in Eq. (8.27) has been evaluated in two ways, (a) from Eq. (8.25) and (b) through regression of measured data from several places; the result obtained from method (a) is $A = 1.97$, and from method (b), $A = 1.91$. On the basis of three different geographic settings, the difference between 1.97 and 1.91 may be considered small. Certainly, more data are needed to further substantiate this equation and for better understanding of the relationships among various parameters as formulated in Eq. (8.25) or equivalents. For details, Eq. (8.25) should be used.

The phenomenon of fumigation near the shoreline as depicted in Fig. 8.8 has been recognized for many years (e.g., Hewson and Olsson, 1967; Lyons and Cole, 1973). A recent study by Deardorff and Willis (1982) showed maximum ground-level concentrations (see Hanna, 1985)

$$\chi_{max} \simeq 0.3 \, Q/(uh^2)$$

which occur at a downwind distance of

$$X_{max} \simeq 10(u/W_*)h$$

where W_* is the convective velocity scale,

$$W_* = (gHh/C_p\rho T)^{1/3}$$

where H is the sensible heat flux and T is the surface temperature.

8.3   Vertical Variations of Wind Speed

Vertical distributions of wind speed in the atmospheric boundary layer are discussed in detail in Chapter 6. However, for engineering applications, *in situ* measurements of the aerodynamic roughness Z_0 are not always available. Many wind-profile laws, such as the simple logarithmic distribution, cannot be applied. Therefore, many engineers have resorted to the power-law wind profile, which, to a large degree, is quite accurate and useful for engineering applications (see, e.g., Panofsky and Dutton, 1984).

The power-law wind profile states that

$$\frac{U_z}{U_1} = \left(\frac{Z}{Z_1}\right)^P \tag{8.31a}$$

where U_z is the wind speed at height Z, U_1 and Z_1 are the wind speed and height already known, respectively, at a reference height, and the exponent P is a function of atmospheric stability and roughness.

Forming the logarithm of Eq. (8.31a),

$$\ln(U_z/U_1) = P \ln(Z/Z_1)$$

Differentiating with respect to height and holding reference (known) parameters as constants, we have

$$\frac{d}{dZ}(\ln U_z - \ln U_1) = P\frac{d}{dZ}(\ln Z - \ln Z_1)$$

or

$$\frac{1}{U_z}\left(\frac{dU_z}{dZ}\right) = \frac{P}{Z}$$

But, from Chapter 6,

$$\frac{dU_z}{dZ} = \frac{U_*}{\kappa Z}\phi_m\left(\frac{Z}{L}\right)$$

Therefore, we get

$$P = \frac{U_*}{\kappa U_z}\phi_m\left(\frac{Z}{L}\right)$$

and setting $C_D = (U_*/U_z)^2$,

$$P = \frac{\sqrt{C_D}}{\kappa}\phi_m\left(\frac{Z}{L}\right) \tag{8.31b}$$

Under near-neutral stability conditions, $\phi_m(Z/L) = 1$, Eq. (8.31b) reduces to

$$P = C_D^{1/2}\kappa^{-1}. \tag{8.31c}$$

An example is shown in Figs. 8.10 and 8.11. As discussed in Chapter 6, C_D over water may be estimated around 1.5×10^{-3} by setting $\kappa = 0.4$, $P = 0.10$. This is in excellent agreement with Eq. (8.31c).

On the basis of many investigations, such as Davenport (1965), Panofsky and Dutton (1984), and Justus (1985), approximate values for the exponent P are summarized in Table 8.5 for practical application in the coastal zone. Note that neutral conditions usually exist when the sky is overcast during the day or night and 1 hr after sunrise and before sunset. Unstable conditions prevail when the sea surface temperature (SST) is higher than the air temperature over

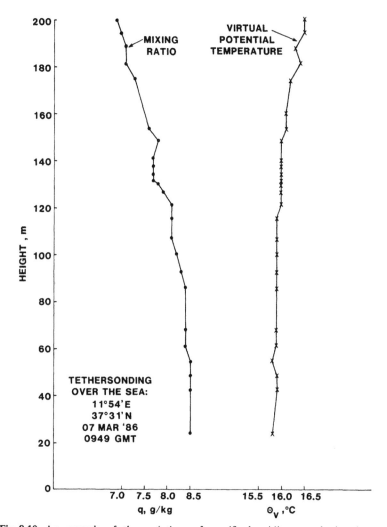

Fig. 8.10. An example of the variations of specific humidity q and virtual potential temperature θ_v over the Mediterranean Sea under near-neutral atmospheric stability conditions on March 7, 1986.

the sea or during daytime on land when the soil temperature is higher than the air temperature. Stable conditions are opposite to unstable conditions—that is, they exist when the sea surface is colder than the air or the soil temperature is lower than the air temperature during the night. Generally speaking, when the air–sea temperature difference, $\Delta\theta$, is within $\pm 2°C$ and when the wind

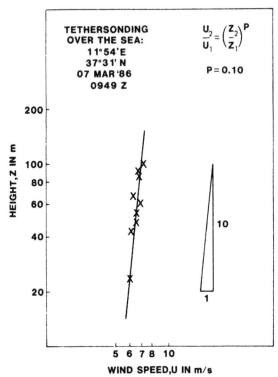

Fig. 8.11. A log–log plot of the wind speed variation with height as measured simultaneously with temperature and humidity, as shown in Fig. 8.10. Note that the exponent P, as indicated in Eq. (8.31a), is 0.10.

Table 8.5

Approximate Values of the Power Law Exponent P
as Functions of Various Coastal Environments
and Atmospheric Stability[a]

	Coastal environments		
Stability	Offshore waters	Flat, open coast	Towns or cities
Unstable	0.06	0.11	0.27
Neutral	0.10	0.16	0.34
Stable	0.27	0.40	0.60

[a] See text for explanation.

speed is higher than 8 m s^{-1} over the sea the stability may be considered neutral (see Hasse and Weber, 1985). This situation exists most often in open sea conditions. Over land when the wind speed is more than 6 m s^{-1} the stability normally is neutral except when the solar altitude is higher than $60°$ [under these conditions, the unstable class should be applied (see Turner, 1969)].

In order to further verify that Table 8.5 is a useful approach in the coastal area, Hsu (1982b) has conducted a study on the flat southern coast of St. Croix, U.S. Virgin Islands. Because of an alumina refinery complex, some villages, and an airport nearby, the coastal environment may be classified as between flat, open coast, and towns and cities (cf. Table 8.5). Our results show that for unstable conditions $P = 0.18$ (see stability B in Hsu, 1982b), which compares very well with $P = 0.19$ as averaged between 0.11 and 0.27 in unstable conditions for the last two categories shown in Table 8.5. Similarly, for neutral stability $P = 0.22$ as compared to $P = 0.25$ (see stability D), and for stable conditions $P = 0.50$ as compared to $P = 0.47$ (see stability F).

Note also that, in the meteorological literature, over smooth open country a value of P close to $\frac{1}{7}$ (or 0.14) is usually found with neutral lapse rates (see, e.g., Blackadar, 1960). Our value of $P = 0.16$ as indicated in Table 8.5 is in agreement with this most commonly quoted value in the literature. On the basis of the above evaluation, Table 8.5 is recommended for general application. However, for detailed computations, Sedefian's (1980) nomogram, which has been further verified by Hsu (1982b) for coastal applications, may be consulted. Because Sedefian's graph needs input of $1/L$, where L is the Monin–Obukhov stability length, one needs further computation from a table given in Justus (1985, Table 33.4) for use on land and from a diagram given by Hasse and Weber (1985, Fig. 3) for use over sea. The value of Z_0 may be obtained from Panofsky and Dutton (1984, Table 6.2 on p. 123).

8.4 Radiometeorology

The performance of microwave and millimeter-wave electromagnetic (EM) systems depends to a large extent on the behavior of the atmosphere. This statement is particularly true in the coastal environment. Examples are shown in Fig. 8.12, which illustrates that an antenna on a low tower near shore might be able to communicate with a ship offshore if it is within the evaporation duct, but might encounter some difficulty in communicating with an oil rig if its receiver is located above the duct. Similarly, if a radar is located on a high tower where its EM rays are trapped in an elevated duct, it would not be able to detect an incoming airplane because of a "radar hole."

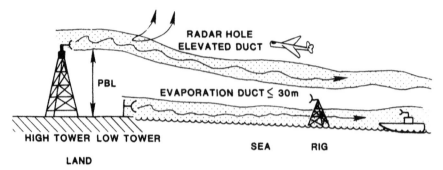

Fig. 8.12. Schematic of radio-ducting problems normally encountered in the coastal zone (see text for explanation).

Over water surfaces, the water vapor decreases rapidly with height because of evaporation. These vapor gradients in turn produce gradients in refractivity N, according to Eq. (8.12), which are large enough to cause the EM rays to be channeled or ducted within approximately 30 m above the surface. This is why the evaporation duct is found regularly over relatively warm bodies of water.

Duct elevation is normally caused by temperature inversion aloft. An inversion can exist because of sinking or subsidence of air mass, such as on the western coasts of the Americas, particularly in areas where cool water upwells from the bottom to the surface. Another example is diurnal warming and cooling of the planetary boundary layer (PBL), that is, the mixed layer depth of mixing height across the coastal zone. Because of the large heat capacity of water, diurnal variation of the PBL is much smaller over a sea than over land. Therefore, large temporal and spatial variability in PBL height exists across the coastal zone. Examples are shown in Fig. 8.13. In order to estimate the height difference, Hsu (1979b) has provided an operational method (Fig. 8.14) that states that during the day

$$H_{land} - H_{sea} \simeq 123(T_{land} - T_{sea})$$

where H (in meters) and T (in degrees Celsius) are the mixing height and surface temperature, respectively. Subscripts land and sea stand for onshore and offshore regions, respectively. The value of H_{land} can be obtained by Weather Service radiosonde data, whereas T is provided by satellites. Thus, H_{sea} can be obtained.

Radio propagation in the atmosphere is affected by variations of the index of refraction. This index is defined as

$$n(h) = c/v$$

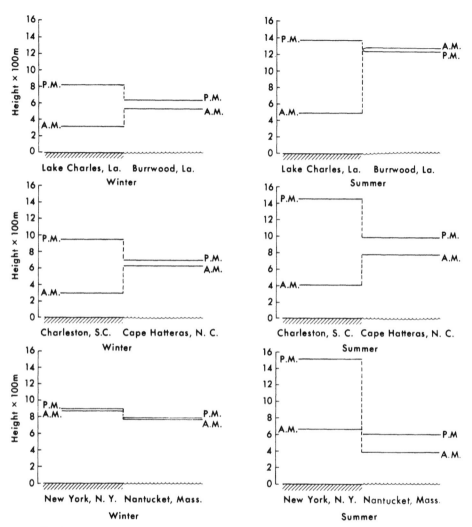

Fig. 8.13. Examples of mean mixing heights in the mornings (a.m.) are compared to those in the afternoon (p.m.) between two approximated stations for land and sea (Hsu, 1979b). Copyright © by D. Reidel Publishing Co. Reprinted by permission.

where c is the velocity of light through a vacuum (3×10^8 m s^{-1}; v is the velocity of light in the air; and h is the height above the earth.

Because n is close to unity (i.e., $n \simeq 1.000300$), it is convenient to define another index, known as radio (or atmospheric) refractivity (N):

$$N(h) = (n - 1) \times 10^6$$

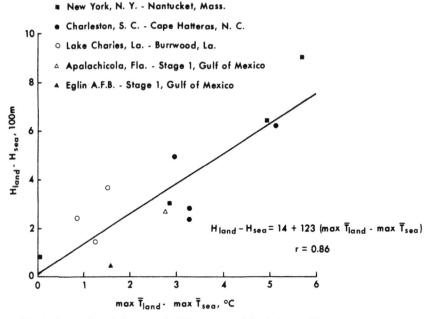

Fig. 8.14. Relationship between the differences in mixing height and in temperature across the coastal zone (Hsu, 1979b). Copyright © by D. Reidel Publishing Co. Reprinted by permission.

For microwave length, N in the troposphere is given by (see, e.g., Bean and Dutton, 1968)

$$n = \frac{77.6}{T}\left(P + \frac{4810e}{T}\right) \tag{8.32}$$

where P is the atmospheric pressure (in millibars), T is the absolute temperature (in degrees Kelvin), and e is the partial pressure of water vapor (in millibars).

Note that the first term on the right-hand side of Eq. (8.32) is often called the dry term and the second, the wet term. Equation (8.32) is considered valid for frequencies ranging from 1 mHz to at least 30 GHz and perhaps up to 72 GHz. For optical propagation, the humidity effect is negligible and N is approximately given by the dry term only (see, e.g., Ishimaru, 1985).

If the refractive index decreases rapidly enough with height, electromagnetic rays may be trapped within a layer of the earth's atmosphere. Such a refractive index profile is usually called a duct (Fig. 8.14). To study wave propagation in ducts, the modified refractive index M is useful:

$$M(h) = N(h) + 10^6 h/a$$

where $a\,(=6378$ km) is the average radius of the earth and h is the height above the ground.

The vertical variation of M with respect to h, that is, dM/dh, may be approximated by

$$\frac{dM}{dh} = \frac{dN}{dh} + \frac{10^6}{a} = \frac{dN}{dh} + 157$$

Therefore, if $dM/dh = 0$, the ray launched horizontally should remain at the same height. If $dM/dh < 0$, the horizontally launched rays will curve downward, as shown in Fig. 8.15a. This is called ground-based duct up to a height h_1, where M is minimum, such as an evaporation duct caused by evaporational cooling of the water surface, that is, warm, dry air on top of cool, moist air. Another type of duct is the elevated duct, shown in Fig. 8.15b. Elevated ducts are the rays that are trapped within the duct between h_1 and h_2. On the other hand, if $dM/dh > 0$, the horizontally launched rays will curve upward.

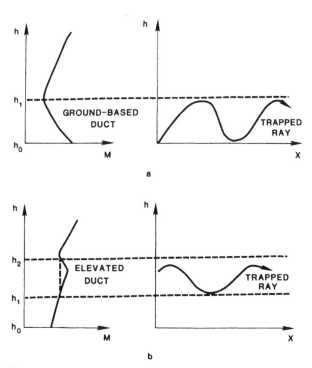

Fig. 8.15. Schematic of M profile and trapped electromagnetic rays.

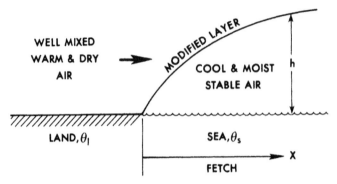

Fig. 8.16. A schematic diagram of the thermally modified boundary layer over a cooler sea caused by advection of warm air (for explanation of symbols, see text).

The above discussion indicates that ducting will occur when the slope dM/dh is negative or if $dN/dh < 157N$ units per kilometer. An example of an elevated duct in the coastal region is shown in Fig. 8.16, in which a thermally modified boundary layer over a cooler sea, caused by advection of warm air, is illustrated. In the figure the height of this modified layer is represented by h; the distance downwind from the coast (i.e., the fetch) is X; and θ_1 and θ_s are the potential temperatures over land and sea, respectively.

Examples of N profile offshore and onshore under these conditions are shown in Fig. 8.17. Note that according to Eq. (8.32), along identical pressure surfaces warm, dry air produces smaller values of N than cool, moist air. It can be seen that in the offshore region in the western Arabian Sea N values near the surface are around 400 units per kilometer, whereas values around 280 were found near the surface over the Empty Quarter on the Arabian Peninsula (Hsu, 1983a). Detailed analyses indicate that the elevated duct layer slopes upward from the shoreline to a height of 620 m at a downwind distance of 960 km. In order to estimate this duct height, Hsu (1983a) suggested that

$$h = cx^{1/2}$$

where h is the elevated duct height in meters over the water, X is the distance downwind from the coast (i.e., the fetch), and c is the proportionality constant. Based on available data as shown in Fig. 8.18, $c \simeq 18$.

Along a tropical desert coast having mountains nearby, microwave communication systems may experience another type of anomalous propagation problem. During the day, hot and dry air is pushed uphill by a sea breeze, but during the night these hot, dry "bubbles" will flow downhill with gravity flow. These hot, dry intrusions can produce radio fading. Some of the

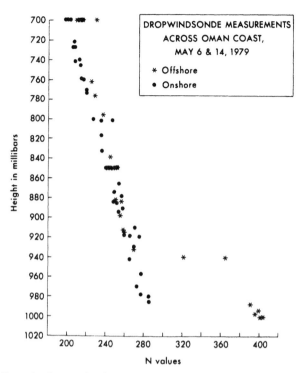

Fig. 8.17. Example of warm, dry air blowing over a cool, moist sea in the western Arabian Sea (Hsu, 1983a).

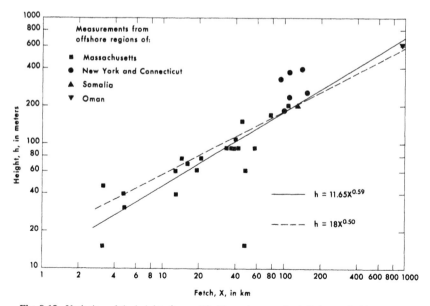

Fig. 8.18. Variation of the height of a modified layer h versus fetch X downwind from the coast owing to warm air advection over cool, moist water, which would produce an elevated duct over the downwind region (Hsu, 1983a).

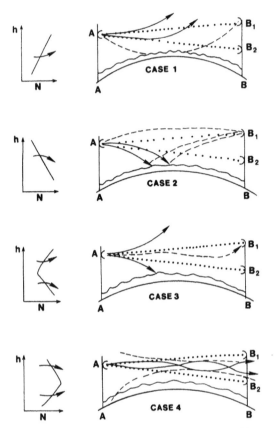

Fig. 8.19. Some aspects of fading mechanisms, which have space and frequency diversities B_1 and B_2. Hot, dry bubbles are illustrated in hatched areas in the figure. Microwave tower A tries to communicate with tower B.

fading mechanisms are illustrated in Fig. 8.19. Case 1 represents subrefraction, under which the rays are bent up. The ray bath is faded by defocusing and diffraction loss. Case 2 is for superrefraction, which is the bending down of rays. Trapping or ducting is the severe bending down of rays. Trapping causes the rays to be guided or ducted by the earth's surface or by other layers or large difference in refractive index. Under superrefractive conditions, the path will be faded by defocusing and multipaths. Case 3 illustrates a dispersive layer. The path is faded by dispersion. Case 4 represents a converging layer. The path is faded by multipath effects and sometimes trapping or ducting.

Atmospheric refractivity undergoes large changes when one sails across an oceanic thermal front. An example is shown in Fig. 8.20. It can be seen that

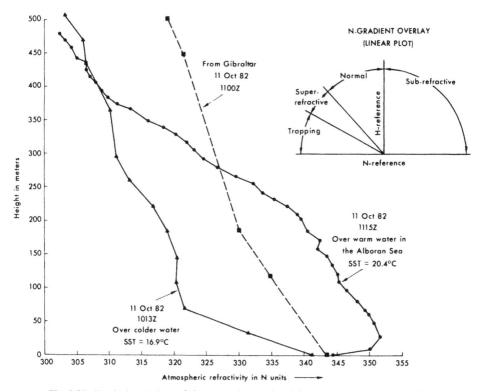

Fig. 8.20. Vertical variations of the atmospheric refractivity across an ocean thermal front in the Alboran Sea (see text for explanation).

over the warm water, multipath fading similar to case 4 in Fig. 8.19 may occur, whereas ducting or trapping may exist over cold water. Figure 8.20 also shows that the radiosonde measurements for a nearby small peninsula such as Gibraltar cannot be used to represent the adjacent marine environment.

8.5 Storm Surges

A storm surge is an abnormal rise of the water level along a shore resulting principally from the atmospheric pressure and the winds of a storm. It is defined as the difference between the actual elevation of the sea surface and the elevation that would have existed, owing to astronomical tide, in the absence of a storm.

A method of estimating storm surge from tropical storms has been developed by Jelesnianski (1972) for coastal regions of the United States. The

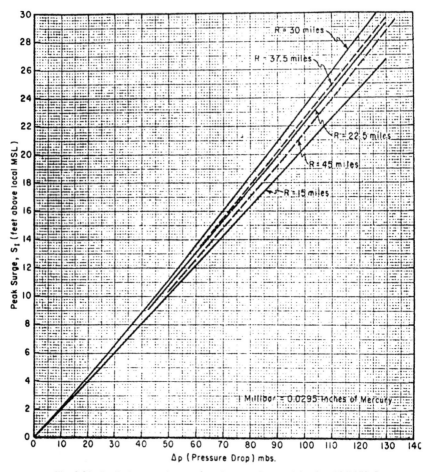

Fig. 8.21. Preliminary estimate of peak surge. [From Jelesnianski (1972).]

corrected peak storm surge S_P is computed as

$$S_P = S_I F_S F_M \tag{8.33}$$

where S_I is the peak open ocean surge, F_S is a shoaling factor, and F_M is a correction factor for storm motion. As shown in Fig. 8.21, S_I is given as a function of the variables ΔP (the drop or deficit in the central pressure) and R (the radius to maximum wind). The storm is assumed to move perpendicularly to a shoreline at a speed of 15 mph (~ 6.7 m s^{-1}). This nomograph assumes that there is a critcal storm size as reflected by the radius of maximum winds, R.

Parameter F_S, as shown in Figs. 8.22 and 8.23, reflects the adjustment for the effect of local bottom topography along the U.S. Gulf and Atlantic coasts, respectively. The maximum surge is further adjusted for the storm's speed V_F and direction of motion ψ relative to the coastline by applying the correction factor for storm motion, F_M, as shown in Fig. 8.24.

An example to illustrate the use of these nomographs is given as follows.

Hurricane Camille, one of the most intense storms in recorded North Atlantic tropical cyclone history, devastated the Mississippi coast with winds up to 190 mph and a surge above 20 ft as its center moved inland on that densely populated coastal strip at about 2200 hr (CDT) on August 17, 1969. According to Bishop (1984), Hurricane Camille went ashore with $\Delta P = 108$ mb and $R = 15.6$ statute miles (~ 25 km). From Fig. 8.21 the peak unadjusted surge $S_1 = 22$ ft (~ 6.7 m) above mean sea level, and from Fig. 8.22, based on the landfall location (west of Biloxi, Mississippi), $F_S = 1.23$. The storm approached the coastline at an angle $\psi = 102°$, with a speed $V_F \simeq 14.5$ mph (~ 6.5 m s^{-1}), so $F_M = 0.96$. The peak surge using the Jelesnianski nomograph method is thus [from Eq. (8.33)]

$$S_P = (22 \quad \text{ft})(1.23)(0.96) = 26.2 \quad \text{ft} \quad (\sim 8 \quad \text{m})$$

The observed peak surge of 24.2 ft (~ 7.4 m) above mean sea level occurred at Pass Christian, Mississippi. This calculated value is in excellent agreement with the value of 8 m at New Orleans, Louisiana, as quoted in Simpson and Riehl (1981, Table 36).

Storm surges in bays and estuaries are more complicated to compute with Eq. (8.33) because of irregular coastline and local topography, as well as the influence of the storm track. Because the circulation around the hurricane is counterclockwise in the northern hemisphere, the wind and waves travel in the same direction on the right-hand side of the storm track, whereas they oppose each other on the left-hand side, as illustrated in Fig. 8.25. Therefore, the storm surge will be higher on the right of the track than on the left. An example is given in Fig. 8.26. Furthermore, according to Goudeau and Conner (1968) as illustrated in Fig. 8.26:

The storm surge associated with a hurricane begins to form while the storm is out at sea. The wave (or waves) then moves shoreward with a speed in shallow water equal to $(gD)^{1/2}$, where g is the acceleration due to gravity and D is the depth of the water. As the wave moves into shallow water near shore it is slowed because of decreasing depth, and its amplitude is increased by shoaling and by convergence of water near shore. The amplitude of the wave may then decrease as it moves inland by the spreading out of water over higher ground, or it may increase by convergence into an inlet or channel. Both amplitude effects were evident in the gauge readings from Betsy.

One example of the amplitude effect was the increasing surge elevation in the Mississippi River to a point (near Baton Rouge, Louisiana, on the upper

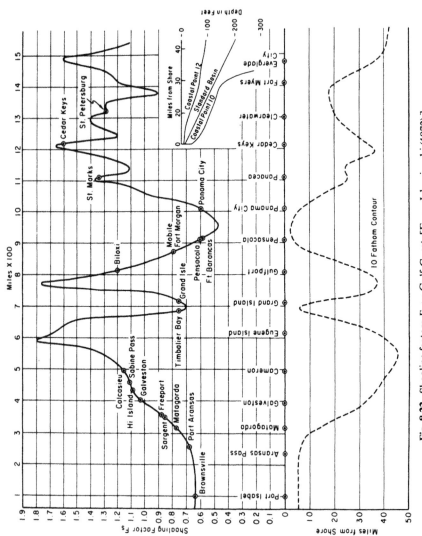

Fig. 8.22. Shoaling factors F_s on Gulf Coast. [From Jelesnianski (1972).]

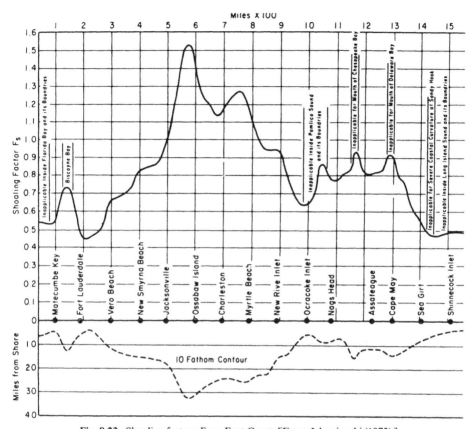

Fig. 8.23. Shoaling factors F_s on East Coast. [From Jelesnianski (1972).]

left corner with 13.9 ft) about 300 km upriver from the mouth. The flooding in the city of New Orleans, Louisiana, to a large extent was caused by the channeling of the water from the southeast over the marsh and up the Gulf Outlet. Overtopping occurred along the levees surrounding both the Industrial Canal and the Intracoastal Waterway. The overtopping caused numerous breaks in the levees surrounding the lower basin. The flood water extended to much of the low ground (below sea level) both west of the Industrial Canal and north of the Gentilly Ridge.

In order to improve flood forecasting, the U.S. National Weather Service has developed two storms surge models. One is called SPLASH (Special Program to List Amplitudes of Surges from Hurricanes), as shown in part in Figs. 8.21–8.24 (Jelesnianski, 1972). Another model is called SLOSH (Sea, Lake, and Overland Surges from Hurricanes), which is an extension of the

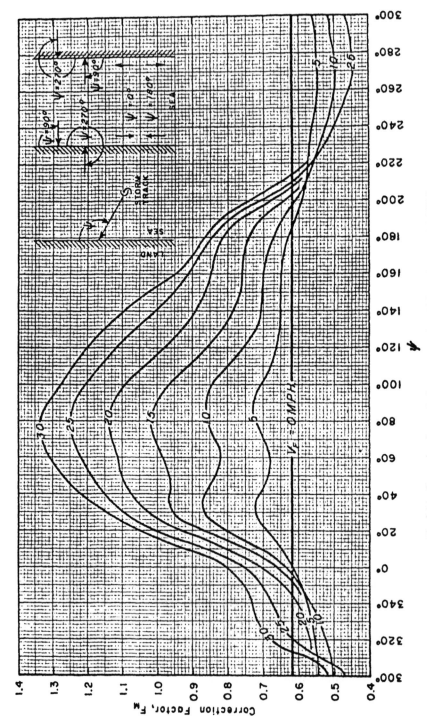

Fig. 8.24. Correction factor F_M for storm motion. [After Jelesnianski (1972).]

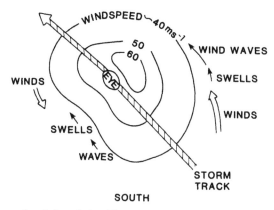

Fig. 8.25. Schematic of the relationship among wind, waves, and hurricane track in the northern hemisphere. Note that the wind and waves travel in the same direction on the right-hand side of the storm track, whereas they travel against each other on the left-hand side.

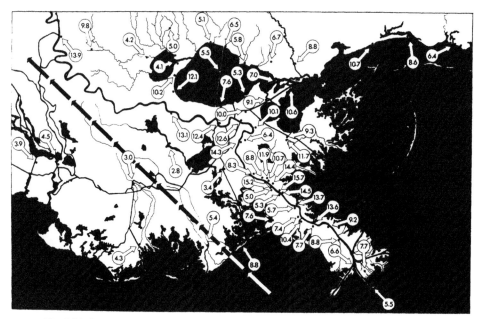

Fig. 8.26. Track of Hurricane Betsy, 1965. High water (in feet) marks Louisiana and Mississippi coastal areas. [After Goudeau and Conner (1968).]

SPLASH model. Crawford (1979) reported an early version of SLOSH that was constructed for the Lake Pontchartrain basin in Louisiana. According to Jarvinen and Lawrence (1985), both

> SLOSH and SPLASH are based on similar principles, but SPLASH storm surge heights are computed only over water and at the coastline (which is modeled as an infinite vertical wall), while SLOSH computations extend inland over the coastal flood plain. Thus, SLOSH results may be used to estimate the inland distribution of flooding and this provides a basis for determining areas for which evacuation is required.

8.6   Estimating Wind Waves and Ocean Currents

8.6.1   Estimating Wind-Generated Waves

When the wind blows over the water surface, waves start to form and grow. The water surface of these wind-generated waves can be described as a superposition of many different forms of waves. The most fundamental description of a two-dimensional, simple progressive wave propagating in the x direction is illustrated in Fig. 8.27. In the figure, the symbol L represents the wave length (the horizontal distance between crest and crest or trough and trough), H is the wave height (the vertical distance to its crest from the preceding trough), T is the wave period (the time required for two successive crests to pass a given point), d is the water depth (the distance between the seabed and the stillwater level), and η is the displacement of the water surface

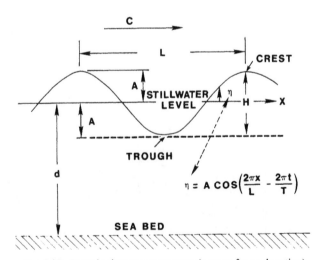

$$\eta = A \cos\left(\frac{2\pi x}{L} - \frac{2\pi t}{T}\right)$$

Fig. 8.27. Some basic wave parameters (see text for explanation).

relative to the stillwater level and is a function of x and time t, according to an equation shown in the figure. At the wave crest, $\eta = A = 0.5H$, where A is the wave amplitude.

The speed at which a wave form propagates is termed the phase velocity or wave celerity C:

$$C = L/T \tag{8.34}$$

An expression relating the wave celerity to the wavelength and water depth is given by (see, e.g., U.S. Army Corps of Engineers, 1984)

$$C = \left[\frac{gL}{2\pi} \tanh\left(\frac{2\pi d}{L} \right) \right]^{1/2} \tag{8.35}$$

where g is the gravitational acceleration and tanh is the hyperbolic tangent.

From Eq. (8.34), Eq. (8.35) can be written as follows:

$$C = \frac{gT}{2\pi} \tanh\left(\frac{2\pi d}{L} \right) \tag{8.36}$$

Gravity waves may be classified by the water depth in which they travel. They are:

(A) For deep water,

$$d/L > \tfrac{1}{2}; \quad 2\pi d/L > \pi; \quad \text{and} \quad \tanh(2\pi d/L) \approx 1$$

Then, from Eq. (8.36),

$$C_0 = gT/2\pi \tag{8.37}$$

where C_0 is the wave celerity in deep water. Note that C_0 is virtually independent of depth.

(B) For shallow water,

$$d/L < \tfrac{1}{25}; \quad 2\pi d/L < \tfrac{1}{4}; \quad \text{and} \quad \tanh(2\pi d/L) \approx 2\pi d/L$$

Then, from Eq. (8.35),

$$C = (gd)^{1/2} \tag{8.38}$$

Thus, when a wave travels in shallow water, wave celerity depends only on water depth.

(C) Between deep and shallow water, d/L ranges from $\tfrac{1}{25}$ to $\tfrac{1}{2}$; $2\pi d/L$ from $\tfrac{1}{4}$ to π; and $\tanh(2\pi d/L)$ does not change. Therefore, Eq. (8.35) or Eq. (8.36) cannot be further simplified.

Although most advanced techniques in wave prediction need numerical models run by sophisticated computers as well as the input of large quantities of meteorological information, the following techniques are presented for field use. They can be employed to estimate probable wave conditions, such as for design studies. Note that wave prediction is called "hindcasting" when based on past meteorological conditions and "forecasting" when based on predicted conditions.

Because the water surface is composed randomly of various kinds of waves with different amplitude, frequency, and direction of propagation, their participation is usually decomposed into many different harmonic components by Fourier analysis so that the wave spectrum can be obtained from a wave record. Various statistical wave parameters can then be calculated. The most widely used parameter is the so-called "significant wave height" $H_{1/3}$, which is defined as the average height of the highest one-third of the waves observed at a specific point. Significant wave height is a particularly useful parameter because it is approximately equal to the wave height that a trained observer would visually estimate for a given sea state (see, e.g., Bishop, 1984).

In order to estimate wave parameters such as $H_{1/3}$ and corresponding period, accurate information on the fetch and duration of the wind is needed. The following "official" procedure is adapted from the U.S. Army Corps of Engineers (1984). Fetch is defined as a region in which the wind speed and direction are reasonably constant. For practical wave predictions it is usually satisfactory to regard the wind speed as reasonably constant if variations do not exceed 5 knots (2.5 m s^{-1}) from the mean. A coastline upwind from the point of interest always limits a fetch. An upwind limit to the fetch may also be provided by curvature or spreading of the isobars or by a definite shift in wind direction. Frequently the discontinuity at a weather front will limit a fetch, although this is not always so. The effect of fetch width on limiting ocean wave growth in a generating area may usually be neglected, since nearly all ocean fetches have widths about as large as their lengths. In inland waters (bays, rivers, lakes, and reservoirs), fetches are limited by landforms surrounding the body of water. Fetches that are long in comparison to width are frequently found. Shorelines are usually irregular, and a more general method for estimating fetch must be applied. A recommended procedure for determining the fetch length consists of constructing nine radials from the point of interest at 3-degree intervals and extending these radials until they first intersect the shoreline. The length of each radial is measured and arithmetically averaged. While 3-degree spacing of the radials is used in this example, any other small angular spacing could be used.

Estimates of the duration of the wind are also needed for wave prediction. Because complete synoptic weather charts are prepared only at 6-hr intervals, interpolation to determine the duration may be necessary. Linear interpolation is adequate for most uses, and, when not obviously incorrect, is usually the best procedure. Care should be taken not to interpolate if short-duration phenomena, such as frontal passage or thunderstorms, are present.

Two simple methods for estimating waves in deep water are given in the Shore Protection Manual (see U.S. Army Corps of Engineers, 1984). One method is based on the parametric (spectral) model originally developed by Hasselmann *et al.* (1976) and the other is based on an empirical–analytical procedure used by Sverdrup and Munk (1947) and later revised by Bretschneider (1952, 1958), the so-called Sverdrup–Munk–Bretschneider (SMB) methods.

The spectral method is outlined as follows.

In the fetch-limited case (i.e., when winds have blown constantly long enough for wave heights at the end of the fetch to reach equilibrium), the parameters required for wave estimates are the fetch F and U_A, the adjusted wind speed (which has been adjusted to standard height of 10 m above the sea surface according to the method outlined in Section 8.3 and in air–sea temperature difference as described in Chapter 6). The spectral wave height H_{mo} and peak spectral period T_m are the parameters predicted.

$$\frac{gH_{mo}}{U_A^2} = 1.6 \times 10^{-3}\left(\frac{gF}{U_A^2}\right)^{1/2} \tag{8.39}$$

$$\frac{gT_m}{U_A} = 2.857 \times 10^{-1}\left(\frac{gF}{U_A^2}\right)^{1/3} \tag{8.40}$$

$$\frac{gt}{U_A} = 6.88 \times 10^{1}\left(\frac{gF}{U_A^2}\right)^{2/3} \tag{8.41}$$

Note that $T_{1/3}$ is given as $0.95T_m$. The preceding equations are valid up to the fully developed wave conditions given by

$$\frac{gH_{mo}}{U_A^2} = 2.433 \times 10^{-1} \tag{8.42}$$

$$\frac{gT_m}{U_A} = 8.134 \tag{8.43}$$

$$\frac{gt}{U_A} = 7.15 \times 10^{4} \tag{8.44}$$

where H_{mo} is the spectrally based significant wave height, T_m is the period of the peak of the wave spectrum, F is the fetch, t is the duration, and U_A is the adjusted wind speed.

The SMB method can be presented best in nomogram form as provided in Fig. 8.28. Note that the U_A line is followed from the left side of the graph across to its intersection with the fetch length or F line, or the duration t line, whichever comes first.

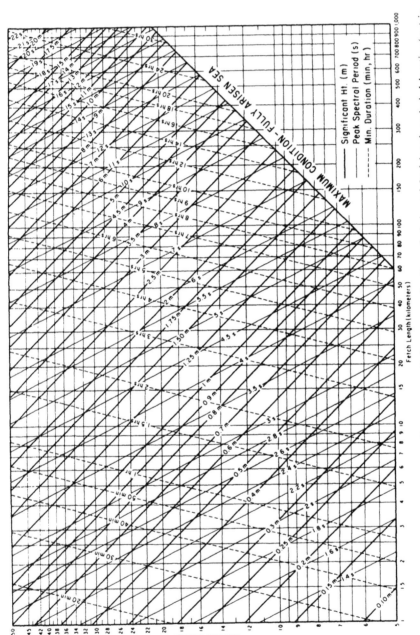

Fig. 8.28. Nomograms of deepwater significant wave prediction curves as functions of wind speed, fetch length, and wind duration (metric units). Note that the ordinate is the adjusted wind speed, which has been corrected for height and atmosphere stability effects (see text for details). [After Shore Protection Manual, U.S. Army Corps of Engineers (1984).]

8.6.2   Estimating Wind-Driven Currents

Under conditions of constant wind, a common method of estimating surface wind-driven currents is the wind factor approach (see, e.g., Bishop, 1984). From Chapter 6, the wind stress τ_{air} can be expressed as

$$\tau_{air} = \rho_{air} C_D U_{air}^2 \tag{8.45}$$

where ρ and U are air density and wind speed, respectively, and C_D is the drag coefficient.

On the other hand, the shear stress in the water, τ_{sea}, is

$$\tau_{sea} = \rho_{sea} C_D U_{sea}^2 \tag{8.46}$$

where ρ_{sea} and U_{sea} are the water density and surface current, respectively.

If the wind has been blowing for several hours so that steady-state condition may be assumed, under which

$$\tau_{air} = \tau_{sea}$$

we have, from Eqs. (8.45) and (8.46),

$$\rho_{air} C_D U_{air}^2 = \rho_{sea} C_D U_{sea}^2$$

or

$$\frac{U_{sea}}{U_{air}} = \left(\frac{\rho_{air}}{\rho_{sea}}\right)^{1/2} \simeq 0.03$$

Therefore, the wind-driven current is approximately 3% of the wind speed measured at about 10 m above the sea surface. The surface current is directed at 20° to the right of the wind in the northern hemisphere and 20° to the left of the wind in the southern hemisphere (Bishop, 1984).

If the wind is not steady, the wind-driven current is said to be transient. James (1966) developed a practical method to estimate these currents. Figure 8.29 shows the James wind-drift forecasting curves. To facilitate the use of the James nomogram, two examples are given (see Bishop, 1984):

1. Assuming a 28-knot wind, which has been blowing for the past 24 hr over a fetch of 100 nautical miles, estimate the wind-driven current. This is done by entering values of 28 knots and 24 hr duration on Fig. 8.29. The current speed is found to be 0.67 knots (34.4 cm s^{-1}). Again, entering values of 28 knots and a fetch of 100 nm, current speed is found to be 0.49 knots (25.2 cm s^{-1}). As in the case of wave forecasting (Fig. 8.28), the smaller fetch-limited quantity is the correct value to use. If the fetch is unknown, use the value found with wind speed and wind duration alone.

2. If a wind blows for 12 hr at 12 knots and then for 12 hr at 24 knots, estimate the current. The procedure is shown in Fig. 8.29. During the first

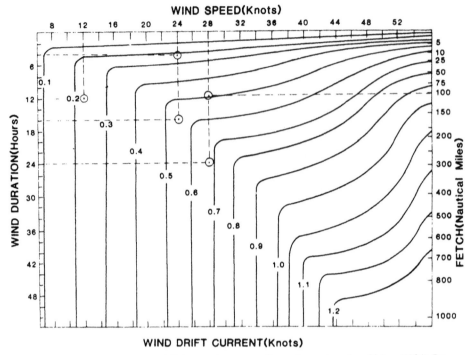

Fig. 8.29. The James wind-drift forecasting curves. Examples are based on Bishop (1984). See text for discussion. [After James (1966).]

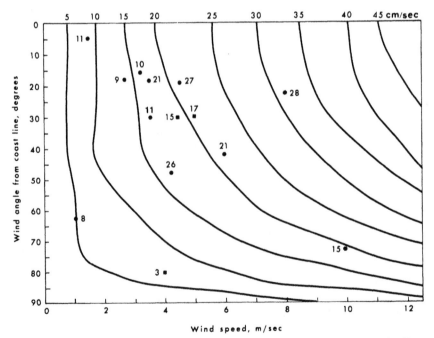

Fig. 8.30. Estimates of wind-driven currents in the nearshore area. For more detail, see Murray (1975).

12 hr, the 12-knot wind generates a current of 0.22 knots. A wind speed of 24 knots could create the same current in 4 hr, a value obtained by entering a wind speed of 24 knots on Fig. 8.29 and reading 4 hr on the left axis. Adding this 4-hr duration to the 12 hr the 24-knot wind has actually blown gives an effective duration of 16 hr. Using 16 hr rather than 12 hr with the 24-knot wind speed gives the correct current speed of 0.54 knots (27.8 cm s^{-1}).

Wind-driven currents in the nearshore area are more complicated. They are strongly controlled by the wind angle to the shoreline and, to a lesser degree, by wind speed (Murray, 1975). For estimating these currents, Fig. 8.30 is provided. Note that there are many other nearshore current systems, such as the rip current, that are beyond the scope of this book, and readers are directed to the Shore Protection Manual (U.S. Army Corps of Engineers, 1984).

8.7 Sand Transport by Wind Action

Sand transport by wind is an important aspect of sedimentological studies of a given region, particularly in coastal areas (e.g., Svasek and Terwindt, 1974). From the perspective of engineering use, a critical and comprehensive literature survey of sand transport by wind on a dry sand surface was made by Horikawa *et al.* (1986). They found that most predictive expressions for the sand transport rate show a cubic dependence on the shear velocity, or obey a similar power law. On the basis of available laboratory and field measurements of the rate of eolian sand transport, Hsu (1971) introduced a relatively simple method by which this rate can be scaled by a special Froude number. Measurements made later by Svasek and Terwindt (1974) further confirm the validity of this method, as shown in Fig. 8.31.

These two approaches may be explained by the following considerations. In high Reynolds number atmospheric turbulence, the budget of turbulent kinetic energy per unit mass may be expressed by (see, e.g., Busch, 1973)

$$\frac{\partial \bar{e}}{\partial t} = \overline{u'w'}\frac{\partial \bar{U}}{\partial z} - \overline{v'w'}\frac{\partial \bar{V}}{\partial z} + \frac{g}{T_0}\overline{\theta'w'} - \varepsilon - \frac{\partial}{\partial z}\left(\overline{w'e} + \frac{1}{\rho_0}\overline{p'w'}\right) \quad (8.47)$$
$$\quad (1) \qquad\qquad (2) \qquad (3) \qquad\quad (4)$$

where $\bar{e} = (\overline{u'^2} + \overline{v'^2} + \overline{w'^2})/2$ and other symbols are given in Chapter 4.

The first terms (1) on the right-hand side of Eq. (8.47) represent the rate at which the turbulent flow, via the Reynolds stresses, extracts energy from the mean flow; it is referred to as the mechanical production term, since it is normally positive. The second term represents the rate at which work is done by the turbulence against buoyancy and thus it is referred to as the thermal production term; the third (3) is the dissipation term; the last terms (4) are the

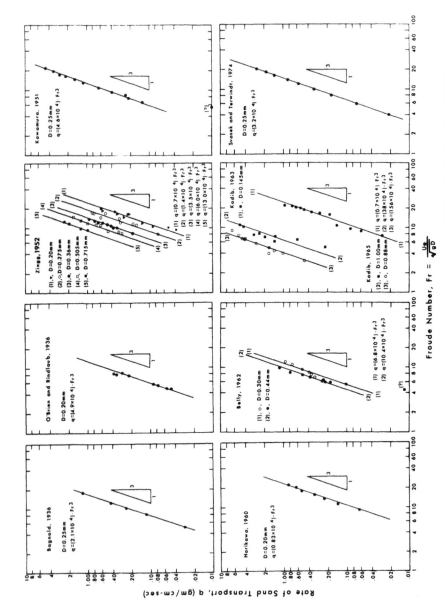

Fig. 8.31. Froude number scaling in the rate of eolian sand transport [cf. Eq. (8.54)]. [See Hsu (1977).]

divergence of the vertical fluxes of turbulent energy (the first of these fluxes is called the turbulent energy flux and the second is the pressure transport).

Under eolian sand transport conditions, we may assume that there is a steady-state flow that is homogeneous over all horizontal planes and that the wind is in the x direction. Equation (8.47) may be written as

$$\frac{\partial \bar{e}}{\partial t} = 0 = -\overline{u'w'}\frac{\partial \bar{U}}{\partial z} + \frac{g}{T_0}\overline{\theta'w'} - \varepsilon - \frac{\partial}{\partial z}\overline{ew'} - \frac{\partial}{\rho_0 \partial z}\overline{p'w'} \qquad (8.48)$$

Furthermore, under the assumptions of constant mean turbulent shear stress and vertical heat flux, Eq. (8.48) may be integrated between the surface and the height Z. This leads to the height-integrated total turbulent kinetic energy equation (Schols and Wartena, 1986)

$$0 = -\overline{u'w'}\,\bar{U} + z\frac{g}{T_0}\overline{w'\theta'} - z\varepsilon_{av} - \overline{ew'} - \frac{1}{\rho_0}\overline{p'w'} \qquad (8.49)$$

The subscript av denotes an average value over the height Z.

On the other hand, transport rate is defined as the total mass of solid particles passing along a lane of unit width in unit time (see, e.g., Schmidt, 1986). The total transport rate is divided into a suspended part, supported by turbulent fluctuations, and the unsuspended, or bedload, portion, which includes particles with rolling, creeping, and saltating motion. Bagnold's (1956, 1966, 1973) development begins with the concept that transport rate is determined by the rate at which work is done by the fluid in moving the particles. According to Schmidt (1986), the total transport rate q, in terms of the immersed weight of the solids, is

$$q = [(\sigma - \rho)/\sigma]mUg \qquad (8.50)$$

where σ is the particle density, m is total mass of solids transported at average wind speed U along unit width of the flow, ρ is fluid density, and g is the gravitational acceleration.

By definition (cf. Chapters 4 and 6).

$$C_z = (U_*/U_z)^2$$

and

$$\tau \equiv -\rho\overline{u'w'} \equiv \rho U_*^2$$

where C_z is the drag coefficient at height z and τ is the wind stress.

Substituting these parameters into Eq. (8.49) and rearranging:

$$\frac{\rho}{g(C_z)^{1/2}}U_*^3 + \frac{\rho z}{T_0}\overline{w'\theta'} - \frac{\rho z\varepsilon_{av}}{g} - \frac{\rho}{g}\overline{ew'} - \frac{1}{g}\overline{p'w'} = 0 \qquad (8.51)$$

where $\rho = \rho_0$.

Since sand transport occurs when wind speed is high, the atmospheric condition may be considered as neutral—that is, the mechanical term is much larger than the buoyancy one (thus the denominator of the Richardson number R_i is much larger than its numerator, making the R_i small). Furthermore, atmospheric measurements showed that under near-neutral conditions, terms such as $\overline{ew'}$ and $\overline{p'w'}$ are all small (see, e.g., Wyngaard and Cote, 1971). Therefore, the energy dissipation term may be considered to be balanced with the mechanical production such that the generalized rate of sand transport is

$$q \propto \frac{\rho}{g} U_*^3 \tag{8.52}$$

where q is in grains per centimeter per second. Many sand transport equations similar to Eq. (8.52) that were formulated directly from a concept similar to that of Eq. (8.52) can now be explained by the consideration of the turbulent energy equation.

On the other hand, Hsu's (1971) approach is based on the concept of nondimensional parameterization, which is a common practice in the study of atmospheric motions such as the formulation of Richardson number on the basis of the ratio of buoyancy to wind shear (see Hess, 1959, p. 290). Returning to Eq. (8.49) and assuming that the diameter D of the sand grain can be lifted at least to a height such that $Z = D$ in order for sand to be airborne, we now rearrange Eq. (8.49) in order to get the nondimensional parameter:

$$0 = \frac{-\overline{u'w'}}{gZ} + \frac{1}{UT_0}\overline{w'\theta'} - \frac{\varepsilon_{av}}{g\bar{U}} - \frac{\overline{ew'}}{g\bar{U}Z} - \frac{\overline{p'w'}}{\rho_0 g\bar{U}Z}$$

As before, under neutral stability ($\overline{w'\theta'} \simeq 0$) and assuming that $\overline{ew'}$ and $\overline{p'w'}$ are small in comparison to the shear production term, we have (by setting $Z = D$ and $U = \bar{U}$)

$$0 = \frac{U_*^2}{gD} - \frac{\varepsilon_{av}}{gU} \tag{8.53}$$

From Eq. (8.50), it is clear that when q is the transport rate per unit mass

$$q \propto gU$$

Therefore, the second term on the right-hand side of Eq. (8.53) represents the rate of sand transport. The nondimensional transport rate can be scaled by the first term of Eq. (8.53). In other words,

$$q = f_1\left(\frac{U_*^2}{gD}\right) = f_2\left(\frac{\tau/\rho}{gD}\right) = f_3\left[\frac{U_*}{(gD)^{1/2}}\right] \equiv f_4(\text{Fr})$$

where $\text{Fr} = U_*/(gD)^{1/2}$ is a special form of the Froude number.

The relationship formulated by Hsu (1971) for computing the rate of sand transport by the wind is

$$q = K(\text{Fr})^3 = K[U_*/(gD)^{1/2}]^3 \qquad (8.54)$$

where q (g cm^{-1} s^{-1}) is the rate of sand transport by the wind and Fr is a special Froude number. The term Fr is a function of the atmospheric shear velocity U_* (cm s^{-1}), the acceleration of gravity g (980 cm s^{-1}), and the mean grain size of the sand particles D (mm). The term K is defined as the dimensional eolian sand transport coefficient and has the same dimensions as q. The values of K are delineated in Fig. 8.32. The value of Fr is explicitly determined by the bracketed term in Eq. (8.54).

If the value of U_* is known, the rate of eolian sand transport can be computed. The following procedures are recommended. In the lowest turbulent layer of the atmosphere over land, under relatively homogenous steady-state conditions, logarithmic horizontal wind velocity increase with height has been observed over beaches (Hsu, 1971; Walters, 1973; Svasek and Terwindt, 1974), over tidal flats (Hsu, 1972b), over deserts (Bagnold, 1954), and in laboratory channels (Kadib, 1965).

The logarithmic wind profile law states that

$$U_z = (U_*/\kappa)[\ln(Z/Z_0)] \qquad (8.55)$$

where U_z is the mean horizontal wind velocity at any given height Z, U_* is the shear (or friction) velocity [equivalent to $(\tau/\rho)^{1/2}$, where τ is the surface wind stress and ρ is the air density], κ is the von Karman constant ($\simeq 0.4$), and Z_0 is the aerodynamic roughness length under the boundary condition that $U_z = 0$ at $Z = Z_0$. The value of Z_0 depends on the characteristics of the underlying surface.

The quantities U_* and Z_0 can be obtained easily from the wind profiles inasmuch as κ/U_* is the slope of the least-square linear fit of the profile on a semilogarithmic scale and Z_0 is the intercept with the ln axis (where $U_z = 0$). Some examples of the logarithmic wind profiles were given by Hsu (1972b) and Svasek and Terwindt (1974).

It is convenient to define (e.g., Priestley, 1959) the drag coefficient C_z at height Z by

$$C_z = \tau/\rho U_z^2 = (U_*/U_z)^2 \qquad (8.56)$$

In other words, if we know C_z for a given area, the shear velocity U_* can be calculated from the wind velocity and then substituted into Eq. (8.54) to give the value of q.

A convenient way to express the shear velocity is the drag coefficient. For reference purposes, C_z at 2 m above the surface is usually employed (e.g., Priestley, 1959). It should be noted that C_z is relatively insensitive to Z_0. An

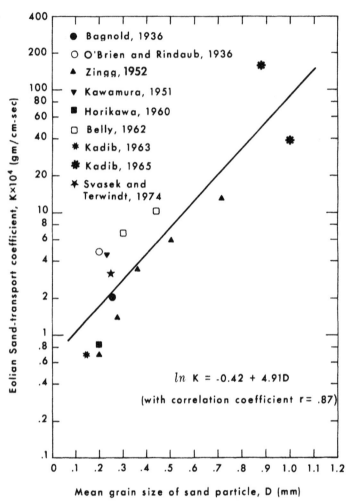

Fig. 8.32. Determination of eolian sand transport coefficient from mean grain size of sand particle. [See Hsu (1977).]

equivalent statement is that the drag coefficient can be specified without having to specify too closely the height to which it relates (Priestley, 1959).

Figure 6.3 summarizes our shear velocity measurements in various coastal environment. Because it is not always convenient for an anemometer to be located 2 m above the surface, the following equation may be used [cf. Eq.(8.55)]:

$$U_* = \kappa U_z / \ln(Z/Z_0) \tag{8.57}$$

where Z_0 may be approximated by applying pertinent values as shown in Fig. 6.3. Note that values of Z_0 were obtained from C_z at 2 m [Fig. 8.32; Eq. (8.55)]. Fortunately, in the atmospheric surface boundary layer [say, Z does not exceed 25 m above the surface, according to Sutton (1953)], the value of U_* may be treated as a constant (Sutton, 1953). Since almost all National Weather Service anemometers are located within this layer, it is safe to use Eq. (8.57) for general micrometeorological applications.

For a sand budget study of a given region, the rate of eolian sand transport is required. The method developed above is based on both field and laboratory experiments (cf. Figs. 8.31 and 8.32), and it should be applicable to natural beaches and deserts such as those discussed by Kadib (1965) and Svasek and Terwindt (1974). Two examples are given next.

1. Inman *et al.* (1966) have stated that $D = 0.15$ mm and $U_{2\,m} \simeq 4$ m s^{-1} (approximate mean annual value) in coastal sand dunes in Baja California, Mexico. From Fig. 8.32, we find that $K = 1.37 \times 10^{-4}$ g cm^{-1} s^{-1}. Since the average relationship between U_* and $U_{2\,m}$ from the scarp through the swale environment [i.e., average of (4), (5), and (7) in Fig. 6.3] indicates that $U_* = 0.07 U_{2\,m}$, we have $U_* = 28$ cm s^{-1}. Substituting g ($= 980$ cm s^{-2}), D ($= 0.015$ cm), K, and U_* into Eq. (8.54), we find that $q = 0.053$ g cm^{-1} s^{-1}. The measured average value for q by Inman *et al.* (1966) is 0.049, which is obtained by applying a mean annual velocity of dune travel (d) of 5 cm day^{-1}, bulk density of the dune sand (ρ_b) of 1.4 g cm^{-3}, and average dune height of 6 m to the equation formulated by Bagnold (1954):

$$q = d\rho_b H/t \tag{8.58}$$

where t is the time of travel (in this case $t = 1$ day or 24×3600 s).

2. Finkel (1959), Lettau (1967), and Lettau and Lettau (1969), among others, studied coastal barchans in southern Peru. Pertinent data used here are based on these publications: $d = 30$ m year^{-1} (or 8.22 cm day^{-1}), $D = 0.314$ mm, $U_{2\,m} \simeq 4$ m s^{-1} (approximate mean annual value; therefore $U_* = 0.07 U_{2\,m} = 28$ cm s^{-1}), $H = 3$ m, $\rho_b = 1.3$ g cm^{-3}. Applying these values into Fig. 8.32 and Eqs. (8.54) and (8.58) as before, we have $q_{predicted} = 0.039$ g cm^{-1} s^{-1} and $q_{observed} = 0.037$ g cm^{-1} s^{-1}. It is evident that the agreement is good.

From these two examples, we conclude that the method and procedure developed in this section are useful for estimating eolian sand transport, not only for natural beaches and deserts but also for coastal sand dunes.

From the above discussions it is clear that U_* is the most important parameter in the computation of eolian sand transport. However, because the beach/dune morphology is not uniform, U_* will vary according to the roughness parameter Z_0. In order to illustrate the coastal environment, a basic flow model for the wind field over an idealized coastal dune is provided in Fig. 8.33.

The development of the basic model of airflow characteristics in the region of coastal sand dunes or ice ridges, as shown in Fig. 8.33, is based on the following considerations.

1. Landsberg (1942) has measured zones of underspeed, overspeed, and cavity corresponding to those of the foredune, dune top, and lee side, respectively, for the sand dunes on the east shore of Lake Michigan.

2. Brooks (1961) reviewed various wind-tunnel experiments dealing with flow over a sudden drop of height H in terrain and over an abrupt change of surface roughness. A cavity is shown to exist for at least a distance of $6.0H$ downwind of the top. Also, a floating layer of high turbulence develops and is still growing in intensity at $X = 6.0H$. The elevation of the turbulent layer at $X = 1.5H$ is the same as that at the original terrain, but at $X = 6.0H$ the turbulent layer has descended to half the previous value. A similar but more detailed study has been conducted by Chang (1966).

3. Inman *et al.* (1966) measured the wind profile over coastal sand dunes in Mexico. The most important finding here was the reversal in wind direction at the toe of the slip face. The reversal indicated a returning flow in the cavity. Also, the air flow was stronger near the top of the dune than in other zones.

4. The size and strength of the lee eddy were studied by Hoyt (1966) on coastal dunes in southwestern Africa and in Georgia (U.S.A.). The controversy over whether there is a lee eddy may now be resolved on the basis of this model (Fig. 8.33). At least two factors play important roles: (a) for dune

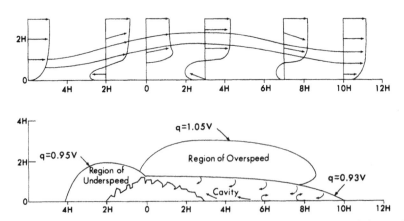

Fig. 8.33. A basic flow model for the wind field in the regions of a coastal sand dune and a similarly shaped ice ridge. In the figure q is the local resultant mean velocity and V is the reference velocity in the uniform stream.

height $H < 5$ m, upstream wind velocity (unaffected by the dune or ridge) must be at least $5-10$ m s^{-1}; for $H > 5$ m, it must be at least $10-20$ m s^{-1}; (b) the spacing between two successive dunes should be larger than the size of the lee eddy (i.e., $5-10H$) in order for the eddy to develop.

5. Laboratory measurements by Rifai and Smith (1971) confirm the field observations.

The model shown in Fig. 8.33, which is substantiated to the degree encompassed by the above discussion, was originally proposed in a more qualitative form by Keeler (1970). It should be noted that this is only a first-approximation verification of the model and calls for refinement. For example, U_* distribution and turbulence intensity measurements and detailed study in regions of underspeed, overspeed, and cavity in relation to the morphological structure of the dunes and ridges themselves would be extremely helpful.

8.8   Transport of Sea Salts in the Shoreline Environment

Atmospheric particles, particularly sea salts, have become an increasingly important subject for investigation in recent years. Sea-salt aerosols are a significant source of condensation nuclei (Dinger *et al.*, 1970). The quantity and quality of sea-salt particles deposited on land may be important in determining the physical and chemical characteristics of coastal soils and plants (El Swaify *et al.*, 1968; James, 1972).

The following description is based mainly on a study of transport of atmospheric sea salt in the coastal zone by Hsu and Whelan (1976). The generation of aerosols depends on many meteorological and oceanographic factors (see, e.g., Roll, 1965). Among those in the coastal region are wind speed, direction, duration, and fetch (which govern the sea state and whitecap distribution), and subaqueous bathymetry (which controls the breaking wave condition in the surf zone) (Fig. 8.34). Therefore, aerosols occur neither as point sources (for whitecaps) nor as line sources (for surf zone) from an atmospheric diffusion viewpoint. The best approach to estimating the aerosol concentration may be to assume a distributed area source, which simply treats the offshore and nearshore area contributions as coming from a continuous distribution of infinitesimal sources.

Because the coastline constitutes a discontinuity in terms of the roughness of the underlying surface, as well as of heat and moisture, the wind must readjust as it passes such areas. The flow does not immediately adapt at all levels to the local surface roughness but does so only in the layer adjacent to the surface. The height of the layer in which the influence of the new roughness

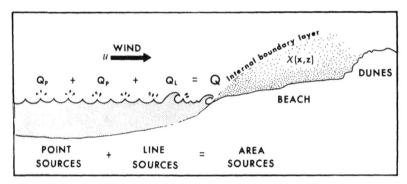

Fig. 8.34. Schematic representation of point and line sources for sea salt and the effect of the internal boundary layer (see text for explanation) (Hsu and Whelan, 1976). Reprinted with permission from the American Chemical Society. .

is felt, called the internal boundary layer (cf. Fig. 8.34 and Section 8.2.4), increases with distance downwind from the point of change in roughness (see, e.g., Blom and Wartena, 1969). Measurements of this boundary layer have been made by Hsu (1971) on a beach and by Panofsky and Peterson (1972) on a narrow peninsula surrounded by bays of varying widths. The thickness of the internal boundary layer is greater under the influence of a sea breeze, owing to stronger solar radiation (Hsu, 1973), than under synoptic onshore winds (e.g., gradient winds). Thus, over land influenced by onshore wind, such as a beach–dune complex, mixing depths, although considerably reduced, are highly variable and act as a sink.

Irregular terrain such as a nearshore cliff or dunes will act as another source of sink, but which condition prevails depends on where the separation of air flow occurs (e.g., Scorer, 1958). Turbulent vortices commonly induced by obstacles should be taken into account. In addition, gravitational settling should be considered as a possible aerosol sink.

For areas where there is no internal boundary layer effect there are many area source formulas for modeling urban air pollution, but a simple and effective one has been developed by Gifford and Hanna (1973). Given the equation

$$\chi = CQ/U \tag{8.59}$$

where χ is the volume concentration of a pollutant emitted from an area source of strength Q and U is the average wind speed, according to their model the parameter C is given by

$$C = (2/\pi)^{1/2} X^{1-b} [a(1 - b)^{-1}] \tag{8.60}$$

where X is the distance from a receptor point to the upwind edge of the sea source. The constants a and b are defined by the vertical atmospheric diffusion length, $\sigma_z = aX^b$. Values of a and b for different atmospheric dispersion conditions can be found in various texts and handbooks (e.g., Environmental Protection Agency, 1973).

For Eq. (8.59) to be useful, it must be verified by available data. Chesselet *et al.* (1972) have measured sea-salt aerosols over the open ocean where $\chi = 2.0$ μg m^{-3} and $U = 5$ m s^{-1}. The value of C may be calculated by Eq. (8.59). For the neutral stability condition, which is prevalent in most oceanic regions (Hsu, 1974b), the value of C ranges from 100 for $X = 5$ m to 133 for $X = 50$ m, assuming $a = 0.080$ and $b = 0.881$ (Environmental Protection Agency, 1973). These values indicate that on the open sea, white-caps would be observable within an upwind distance of 5–50 m. This assumption is reasonable because on fully developed seas, whitecaps are distributed in such a way that the most common distance between two caps ranges from 10 to 100 m (Neumann and Pierson, 1966). Substituting values of χ, U, and C into Eq. (8.59), $Q \simeq 0.1$ μg m^{-2} s^{-1}.

The validity of Q may be examined from another viewpoint. Since the production of atmospheric sea salts from the ocean surface must equal the input from both precipitation and dry fallout, one can follow the data provided by Eriksson (1959) for calculating the input of sea salt into the ocean. If Q is a reasonable number, then the annual input should equal the production as calculated from Q. Given $Q = 0.1$ μg m^{-2} s^{-1} and that the area of the ocean surface is 3.6×10^{14} m^2, then the annual production of sea salts is 1.1×10^5 g year^{-1}. If the fallout rate [from data provided by Eriksson (1959) from two trade-wind stations (6 m s^{-1})] is 5.5×10^{-6} μg cm^{-2} s^{-1} or 540×10^6 tons year^{-1}, and assuming that sea salts are equally removed by dry fallout and precipitation, then the removal rate of sea salts from the atmosphere to the oceans is 1.8×10^{15} g year^{-1}. Thus, Q appears to be a valid number on the basis of the geochemical balance of the world's oceans.

We now examine the estimation of vertical distribution of aerosols in the coastal zone. First, because atmospheric stability in the coastal area plays an important role in the vertical structure of horizontal winds, for a given beach (Hsu, 1973) whose roughness parameter is known (Hsu, 1974c), the power-law relationship with stability characteristics (e.g., Environmental Protection Agency, 1973) is adapted:

$$U/U_1 = (Z/Z_1)^P \tag{8.61}$$

where U_1 is the wind speed at some reference level Z_1. The exponent P depends on the stability class. Substituting Eq. (8.61) into Eq. (8.59),

$$\chi = CQ/[U_1(Z/Z_1)^P] \tag{8.62}$$

As discussed previously, the internal boundary layer, gravitational settling, and terrain-induced vortices will act as a source or as a sink, depending upon the aerodynamic roughness over the area in question. Equation (8.62) should be modified as follows:

$$\chi = (CQZ_1/U_1)Z^{-P} \pm S \tag{8.63}$$

where S represents source (positive) and sink (negative).

For a given location on the beach, the values of C, Q, U_1, and Z_1 can be estimated from known meteorological and oceanographic conditions. Therefore, the vertical distribution from a given region (such as a beach face) toward the open ocean is

$$\chi = \pm S + BZ^{-1/4} \tag{8.64}$$

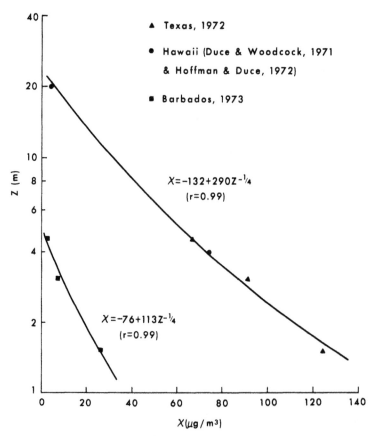

Fig. 8.35. Model verification by available data obtained on open coasts in Texas and Hawaii and on a protected coast on Barbados. Note that the tower top in Duce and Woodcock (1971) was 24 m above sea level and that of Hoffman and Duce (1972) was 20 m above sea level (Hsu and Whelan, 1976). Reprinted with permission from the American Chemical Society.

where $B = CQZ_1^{1/4}/U_1$ from Eq. (8.63), in which $\frac{1}{4}$ is substituted for P to represent neutral stability conditions of an average day and night (Environmental Protection Agency, 1973).

Figure 8.35, in which data from Texas, Hawaii, and Barbados are incorporated, accurately supports Eq. (8.64). Note that the concentration downwind from a pollution source should also depend on the obliqueness of the wind. However, a numerical experiment by Calder (1973) shows that, although the concentrations increase as the wind more closely parallels the direction of a line source, the increase is only slight and, for many practical purposes, may be disregarded.

Appendix A | Units, Constants, and Conversions

The units used here are in meter–kilogram–second (MKS) or the Systeme International (SI).

Basic Units

Length (meters, m)
  1 m = 100 cm = 3.281 ft = 39.37 in
Mass (kilograms, kg)
  1 kg = 10^3 g = 2.205 lb
Time (seconds, s)
Temperature (degrees Kelvin, K)
  $K = °C + 273.16 = \frac{5}{9}(°F - 32) + 273.16$
  (where °C stands for degrees Celsius and °F represents degrees Fahrenheit)

Derived Units

Velocity (m s^{-1})
  1 m s^{-1} = 2.24 mph = 1.94 knots
  Note: 1 m s$^{-1} \approx$ 2 knots
Force (newton, N, kg m s^{-2})
  or dyne, g cm s^{-2})
  1 N = 10^5 dyn = 0.225 lb-force
Work, energy (joule, J; Nm, kg m^2 s^{-2};
  or dyne cm, erg)

$1 \text{ J} = 10^7 \text{ erg} = 0.738 \text{ ft-lb} = 2.778 \times 10^{-7} \text{ kW-hr}$

Note: For heat energy, 1 gram-calorie $= 4.187 \text{ J}$; 1 Btu $= 252.0 \text{ cal}$
$= 1055 \text{ J}$

Power (watt, W; J s^{-1}, $\text{kg m}^2 \text{ s}^{-3}$)

$1 \text{ W} = 10^7 \text{ erg/sec}$

Note: $1 \text{ hp} = 746 \text{ W}$

For power per unit area (W m^{-2})

$1 \text{ W m}^{-2} = 10^3 \text{ erg s}^{-1} \text{ cm}^{-2}$

Also, 1 ly (langley) $= 697.8 \text{ W m}^{-2}$
$= 1 \text{ cal min}^{-1} \text{ cm}^{-2}$

Pressure [force per area, Pascal (Pa), N m^{-2}]

$1 \text{ N m}^{-2} = 10 \text{ dyn cm}^{-2} = 10^{-2} \text{ mb} = 10^{-5} \text{ bar}$

Note: $1 \text{ atm} = 1.013 \times 10^5 \text{ N m}^{-2} = 1013 \text{ mb} = 760 \text{ mmHg}$
$= 29.92 \text{ in Hg} = 14.69 \text{ lb in}^{-2}$

Constants

Stefan–Boltzmann constant (σ)	$5.67 \times 10^{-8} \text{ W m}^{-2} \text{ K}^{-4}$
Universal gas constant ($R*$)	$8314 \text{ J K}^{-1} \text{ kmol}^{-1}$
Speed of light in vacuum (c)	$2.998 \times 10^8 \text{ m s}^{-1}$
Gravitational acceleration (g)	9.81 m s^{-2}
Angular velocity of earth rotation (Ω)	$2\pi \text{ rad day}^{-1}$
	or $7.292 \times 10^{-5} \text{ s}^{-1}$
Molecular weight of dry air (M_d)	28.97
Gas constant of dry air (R_d)	$287 \text{ J kg}^{-1} \text{ K}^{-1}$
Specific heat at constant pressure for dry air (C_p)	$1004 \text{ J kg}^{-1} \text{ K}^{-1}$
Specific heat at constant volume for dry air (C_v)	$717 \text{ J kg}^{-1} \text{ K}^{-1}$
Density of dry air at 0°C and 1000 mb (ρ)	1.28 kg m^{-3}
Density of dry air at 30°C and 1000 mb (ρ)	1.15 kg m^{-3}
Molecular weight of moist air (H_2O in M_v)	18.0
Latent heat of vaporization at 0°C	$2.500 \times 10^6 \text{ J kg}^{-1}$
Latent heat of vaporization at 40°C	$2.406 \times 10^6 \text{ J kg}^{-1}$

Appendix B | The Beaufort Wind Scale

The Beaufort wind scale is a numerical scale devised in 1808 by Admiral Sir Francis Beaufort of the British Navy. It is a system of estimating and reporting wind speeds (forces) for maritime operations. The Beaufort wind scale has been adopted for both land and the open sea by the World Meteorological Organization. Specifications are given in Table B.1.

Table B.1

Beaufort Wind Scale and Specifications

Beaufort number	Description	Wind speed				Specifications on land	Specifications at sea
		m s⁻¹ ᵃ	mph	knots			
0	Calm	<0.6	<1	<1		Calm; smoke rises vertically	Sea like a mirror
1	Light air	0.7–2.3	1–3	1–3		Direction of wind shown by smoke drift but not by wind vanes	Ripples with the appearance of scales are formed, but without foam crests
2	Light breeze	2.4–4.4	4–7	4–6		Wind felt on face; leaves rustle; ordinary vanes moved by wind	Small wavelets, still short but more pronounced; crests have a glassy appearance and do not break
3	Gentle breeze	4.5–6.6	8–12	7–10		Leaves and small twigs in constant motion; wind extends light flag	Large wavelets; crests begin to break; foam of glassy appearance; perhaps scattered white horses
4	Moderate breeze	6.7–8.9	13–18	11–16		Raises dust and loose paper; small branches are moved	Small waves, becoming longer; fairly frequent white horses
5	Fresh breeze	9.0–11.3	19–24	17–21		Small trees in leaf begin to sway; crested wavelets form on inland waters	Moderate waves, taking a more pronounced long form; many white horses are formed (chance of some spray)
6	Strong breeze	11.4–13.8	25–31	22–27		Large branches in motion; whistling heard in telegraph wires; umbrellas used with difficulty	Large waves begin to form; the white foam crests are more extensive everywhere (probably some spray)
7	Near gale	13.9–16.4	32–38	28–33		Whole trees in motion; inconvenience felt when walking against wind	Sea heaps up and white foam from breaking waves begins to be blown in streaks along the direction of the wind
8	Gale	16.5–19.0	39–46	34–40		Breaks twigs off trees; generally impedes progress	Moderately high waves of greater length; edges of crests begin to break into spindrift; foam is blown in well-marked streaks along the direction of the wind

(Continues)

Table B.1 (*continued*)

Beaufort number	Description	Wind speed			Specifications on land	Specifications at sea
		m s^{-1} [a]	mph	knots		
9	Strong gale	19.1–21.8	47–54	41–47	Slight structural damage occurs (chimney pots and slates removed)	High waves; dense streaks of foam along the direction of the wind; crests of waves begin to topple, tumble, and roll over; spray may affect visibility
10	Storm	21.9–24.8	55–63	48–55	Seldom experienced inland; trees uprooted; considerable structural damage	Very high waves with long overhanging crests; the resulting foam, in great patches, is blown in dense white streaks along the direction of the wind; on the whole, the surface of the sea takes a white appearance; the tumbling of the sea becomes heavy and shock-like; visibility affected
11	Violent storm	24.9–28.2	64–72	56–63	Very rarely experienced; widespread damage	Exceptionally high waves (small and medium-sized ships might be for a time lost to view behind the waves); the sea is completely covered with long white patches of foam lying along the direction of the wind; everywhere the edges of the wave crests are blown into froth; visibility affected
12	Hurricane	> 28.2	≥ 73	≥ 64	Not specified	The air is filled with foam and spray; sea completely white with driving spray; visibility very seriously affected

[a] From Roll (1965).

Appendix C | The Saffir/Simpson Damage-Potential Scale

The Saffir/Simpson Damage-Potential Scale (see Simpson and Riehl, 1981) is used by the U.S. National Weather Service to give public-safety officials a continuing assessment of the potential for wind and storm-surge damage from a hurricane in progress. Scale numbers are made available to public-safety officials when a hurricane is within 72 hr of landfall.

The damage-potential scale indicates probable property damage and evacuation recommendations as listed below according to the scale numbers in Table C.1:

1. Winds of 74–95 mph (33–42 m s^{-1}). Damage primarily to shrubbery, trees, foliage, and unanchored mobile homes. No real damage to other structures. Some damage to poorly constructed signs. And/or storm surge 4–5 feet (~1.5 m) above normal. Low-lying coastal roads inundated, minor pier damage, some small craft torn from moorings in exposed anchorage.

2. Winds of 96–110 mph (43–49 m s^{-1}). Considerable damage to shrubbery and tree foliage; some trees blown down. Major damage to exposed mobile homes. Extensive damage to poorly constructed signs. Some damage

Table C.1

Saffir/Simpson Damage-Potential Scale Ranges

Scale number (category)	Central pressure		Winds (mph)	Surge (ft)	Damage
	Millibars	Inches			
1	≥980	≥28.94	74–95	4–5	Minimal
2	965–979	28.50–28.91	96–110	6–8	Moderate
3	945–964	27.91–28.47	111–130	9–12	Extensive
4	920–944	27.17–27.88	131–155	13–18	Extreme
5	<920	<27.17	>155	>18	Catastrophic

to roofing materials of buildings; some window and door damage. No major damage to buildings. And/or storm surge 6–8 ft ($\sim$2–2.5 m) above normal. Coastal roads and low-lying escape routes inland cut by rising water 2–4 hr before arrival of hurricane center. Considerable damage to piers; marinas flooded. Small craft torn from moorings in unprotected anchorages. Evacuation of some shoreline residences and low-lying island areas required.

3. Winds of 111–130 mph (50–58 m s^{-1}). Foliage torn from trees; large trees blown down. Some damage to roofing materials of buildings; some window and door damage. Some structural damage to small buildings. Mobile homes destroyed. And/or storm surge 9–12 ft ($\sim$2.6–3.9 m) above normal. Serious flooding at coast; many smaller structures near coast destroyed; larger structures near coast damaged by battering waves and floating debris. Low-lying escape routes inland cut by rising water 3–5 hr before hurricane center arrives. Flat terrain 5 ft (1.5 m) or less above sea level flooded inland 8 miles ($\sim$13 km) or more. Evacuation of low-lying residences within several blocks of shoreline possibly required.

4. Winds of 131–155 mph (59–69 m s^{-1}). Shrubs and trees blown down; all signs down. Extensive damage to roofing materials, windows, and doors. Complete failure of roofs on many small residences. Complete destruction of mobile homes. And/or storm surge 13–18 ft ($\sim$4–5.5 m) above normal. Flat terrain 10 ft ($\sim$3 m) or less above sea level flooded inland as far as 6 miles ($\sim$10 km). Major damage to lower floors of structures near shore due to flooding and battering by waves and floating debris. Low-lying escape routes inland cut by rising water 3–5 hr before hurricane center arrives. Major erosion of beaches. Massive evacuation of all residences within 500 yards ($\sim$455 m) of shore possibly required, and evacuation of single-story residences on low ground within 2 miles ($\sim$3 km) of shore required.

5. Winds greater than 155 mph (69 m s^{-1}). Shrubs and trees blown down; considerable damage to roofs of buildings; all signs down. Very severe and extensive damage to windows and doors. Complete failure of roofs on many residences and industrial buildings; extensive shattering of glass in windows and doors. Some complete building failures. Small buildings overturned or blown away. Complete destruction of mobile homes. And/or storm surge greater than 18 ft ($\sim$5.5 m) above normal. Major damage to lower floors of all structures less than 15 ft ($\sim$4.5 m) above sea level within 500 yards ($\sim$455 m) of shore. Low-lying escape routes inland cut by rising water 3–5 hr before hurricane center arrives. Massive evacuation of residential areas on low ground within 5–10 miles ($\sim$8 to 16 km) of shore possibly required.

Appendix D | Decomposition of the Vector Wind into *u* and *v* Components

Because the wind is a vector quantity with magnitude C and direction α, it is useful to decompose the vector wind into a scalar west wind (westerly) component and a southerly component v. Note that α represents the direction from which the wind is blowing and blows from north (0° or 360°) to east (90°) to south (180°) to west (270°) and back to north (360°). Thus it is different from the regular cartesian (trigonometric) coordinate system. Note also that either or both u and v can be positive or negative, depending on α. Figure D.1

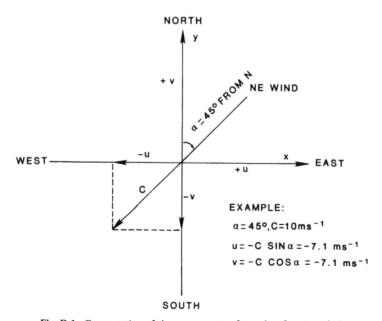

Fig. D.1. Computation of the components of *u* and *v* of vector wind.

illustrates this (see, e.g., Saucier, 1955):

$$u = -C \sin \alpha \qquad v = -C \cos \alpha$$

where positive u is the westerly and positive v is the southerly components of the vector wind C with magnitude C and direction α at a given point in space.

Appendix E | List of Symbols for Surface Analyses[a]

COLOR	SYMBOL	DESCRIPTION
Blue	**H**	High pressure center
Red	**L**	Low pressure center
Blue		Cold front
Blue		Cold front aloft
Red		Warm front
Red/Blue		Stationary front
Purple		Occluded front
Blue		Cold frontogenesis
Red		Warm frontogenesis
Red/Blue		Stationary frontogenesis
Blue		Cold frontolysis
Red		Warm frontolysis
Red/Blue		Stationary frontolysis
Purple		Occluded frontolysis
Purple		Squall Line
Brown		Trough
Yellow		Ridge

[a] Colors are those suggested for on-station use.

Source: Aviation Weather Services: A Supplement to *Aviation Weather AC-006A*, 1975, available through Federal Aviation Administration, U.S. Department of Transportation, Washington, D.C.

References

Amorocho, J., and DeVries, J. J. (1981). Reply. *J. Geophys. Res.* **86**, 4308.

Anthes, R. A. (1982). "Tropical Cyclones—Their Evolution, Structure and Effects." American Meteorological Society, Boston.

Ardanuy, P. (1979). On the observed diurnal oscillation of the Somali jet. *Mon. Weather Rev.* **107**, 1694–1700.

Assaf, G., and Kessler, J. (1976). Climate and energy exchange in the Gulf of Aqaba (Eilat). *Mon. Weather Rev.* **104**, 381–385.

Atkinson, B. W. (1981). "Meso-scale Atmospheric Circulations." Academic Press, New York.

Atkinson, G. D. (1971). "Forecasters' Guide to Tropical Meteorology." Air Weather Service, U.S. Air Force.

Bagnold, R. A. (1954). "The Physics of Blown Sand and Desert Dunes." Methuen, London.

Bagnold, R. A. (1956). The flow of cohesionless grains in fluids. *Philos. Trans. R. Soc. London, Ser. A* **249**, 235–297.

Bagnold, R. A. (1966). An approach to the sediment transport problem from general physics. U.S. Geological Survey Professional Paper No. 411.

Bagnold, R. A. (1973). The nature of saltation and of "bed load" transport in water. *Philos. Trans. R. Soc. London, Ser. A* **332**, 472–504.

Bannon, P. R. (1979a). On the dynamics of the East African Jet. I. Simulation of mean conditions for July. *J. Atmos. Sci.* 36, 2139–2152.

Bannon, P. R. (1979b). On the dynamics of the East African Jet. II. Jet transients. *J. Atmos. Sci.* **36**, 2153–2168.

Bean, B. R., and Dutton, E. J. (1968). "Radio Meteorology." Dover, New York.

Bishop, J. M. (1984). "Applied Oceanography." Wiley, New York.

Blackadar, A. K. (1960). A survey of wind characteristics below 1500 ft. *Meteorol. Monogr.* **4**, 22, 3–11.

Blanc, T. V. (1985). Variation of bulk-derived surface flux, stability, and roughness results due to the use of different transfer coefficient schemes. *J. Phys. Oceanogr.* **15**, 659–669.

Blom, J., and Wartena, L. (1969). The influence of changes in surface roughness on the development of the turbulent boundary layer in the lower layers of the atmosphere. *J. Atmos. Sci.* **26**, 255–265.

Bosart, L. F. (1981). The Presidents' Day snowstorm of 18–19 February 1979: A subsynoptic-scale event. *Mon. Weather Rev.* **109**, 1542–1566.

Bretschneider, C. L. (1952). Revised wave forecasting relationships. *Proc. Conf. Coastal Eng.*, 2nd, *ASCE, Counc. Wave Res.*

Bretschneider C. L. (1958). Revisions in wave forecasting: Deep and shallow water. *Proc. Conf. Coastal Eng.*, 6th, *ASCE, Counc. Wave Res.*

Briggs, G. A. (1973). Diffusion estimation for small emissions. ATDL Contribution File No. 79, Oak Ridge, TN: ATDL/NOAA.

Brooks, F. A. (1961). Need for measuring horizontal gradients in determining vertical eddy transfers of heat and moisture. *J. Meteorol.* **18**, 589–596.

Brutsaert, W. H. (1982). "Evaporation into the Atmosphere." Riedel, Dordrecht.

Budyko, M. I. (1956). Teplovoi Balans Zemnoi Poverkhnosti. Gidrometeorologicheskoe Izdatel'stvo, Leningrad. (trans.: N. A. Stepanova, 1958: "The Heat Balance of the Earth's Surface." Office of Technical Services, U.S. Dept. Commerce, Washington, D.C.

Busch, N. E. (1973). On the mechanics of atmospheric turbulence. *In* "Workshop on Micrometeorology" (D. A. Haugen, ed.), pp. 1–65. American Meteorological Society, Boston.

Businger, J. A. (1973). Turbulence transfer in the atmospheric surface layer. *In* "Workshop on Micrometeorology" (D. A. Haugen, ed.), pp. 67–100. American Meteorological Society, Boston.

Businger, J. A., and Seguin, W. (1977). "Sea-air Surface Fluxes of Latent and Sensible Heat and Momentum," p. 441. U.S. GATE Central Program Workshop, National Center for Atmospheric Research, Boulder, Colorado.

Byers, H. R. (1974). "General Meteorology," 4th Ed. McGraw-Hill, New York.

Calder, K. L. (1973). On estimating air pollution concentrations from a highway in an oblique wind. *Atmos. Environ.* **7**, 863–868.

Chang, S. C. (1966). Velocity distributions in the separated flow behind a wedge-shaped model hill. M.S. thesis, Colorado State University.

Charnock, H. (1955). Wind stress on a water surface. *Q. J. R. Meteorol. Soc.* **81**, 639–640.

Chesselet, R., Morelli, J., and Burt-Menard, P. (1972). Variations in ionic ratios between reference sea water and marine aerosols. *J. Geophys. Res.* **77**, 5116–5131.

Coantic, M. F. (1978). Coupled energy transfer and transformation mechanisms across the ocean atmosphere interface. *Proc. Int. Heat Transfer Conf.*, *6th, Natl. Res. Counc. Can.* **6**, 73–87.

Colon, J. A. (1963). Seasonal variations in heat flux from the sea surface to the atmosphere over the Caribbean Sea. *J. Geophys. Res.* **68**, 1421–1430.

Crawford, K. (1979). Hurricane surge potentials over southeast Louisiana as revealed by a storm surge forecast model: A preliminary study. *Bull. Am. Meteorol. Soc.* **60**, 442–429.

Crisp, C. A. (1979). Training guide for severe weather forecasters. AFGWC/TN-79/002, Air Weather Service, U.S. Air Force.

Danard, M. (1981). A note on estimating the height of the constant flux layer. *Boundary-Layer Meteorol.* **20**, 397–398.

Davenport, A. G. (1965). The relationship of wind structure to wind loading. *Nat Phys. Lab. Symp.* (16), 54–102.

De Angelis, D. (1980). Hurricane alley. *Mar. Weather Log* **24**, 24–28.

Deardorff, J. W., and Willis, G. E. (1967). The free-convection temperature profile. *Q. J. R. Meteorol. Soc.* **93**, 166–175.

Deardorff, J. W., and Willis, G. E. (1982). Ground level concentrations due to fumigation into an entraining mixing layer. *Atmos. Environ.* **16**, 1159–1170.

DeLeonibus, P. S., and Simpson, L. S. (1972). Case study of duration-limited wave spectra observed at an open-ocean tower. *J. Geophys. Res.* **77**, 4555–4569.

Dinger, J. E., Howell, H. B., and Wajciechowski, T. A. (1970). On the source and composition of cloud nuclei in a subsident air mass over the North Atlantic. *J. Atmos. Sci.* **27**, 791–797.

Donelan, M. A. (1982). The dependence of the aerodynamic drag coefficient on wave parameters. *Preprint Conf. Meteorol. Air-Sea Interact. Coastal Zone, The Hague* pp. 381–387.

Driedonks, A. G. M. (1982). Sensitivity analysis of the equations for a convective mixed layer. *Boundary-Layer Meteorol.* **22**, 475–480.

Druilhet, A., Herrada, A., Pages, J. P., and Saissac, J. (1982). Etude expérimentale de la couche limite interne associée à la brize de mer. *Boundary-Layer Meteorol.* **22**, 511–524.

Duce, R. A., and Woodcock, A. H. (1971). Difference in chemical composition of atmospheric sea salt particles produced in the surf zone and on the open sea in Hawaii. *Tellus* **23**, 427–434.

Dunn, G. E., and Miller, B. I. (1960). "Atlantic Hurricanes." Louisiana State Univ. Press, Baton Rouge.

El Swaify, S. A., Swindale, L. D., and Uehara, G. (1968). Salinity tolerances of certain tropical soils and relationships between sodium ion activities and soil physical properties. *Hawaii Inst. Geophys., Univ. Hawaii* **GH-68-12**.

Environmental Protection Agency (1973). "User's Guide for the Climatological Dispersion Model." Natl. Environ. Res. Center, Research Triangle Park, N. Carolina.

Eriksson, E. (1959). The yearly circulation of chloride and sulfur in nature, meteorological, geochemical, and pedological implications. *Tellus* **11**, 375–403.

Estoque, M. A. (1962). The sea breeze as a function of the prevailing situation. *J. Atmos. Sci.* **19**, 244–250.

Fett, R. W. (1979). "Environmental Phenomena and Effects." Navy Tactical Applications Guide, Vol. 2. Naval Environmental Prediction Research Facility, Monterey, California.

Finkel, H. J. (1959). The barchans of southern Peru. *J. Geol.* **67**, 614–647.

Fisher, E. L. (1960). An observational study of the sea breeze. *J. Meteorol.* **17**, 645–660.

Flohn, H. (1965). Studies on the meteorology of tropical Africa. *Meteorol. Inst., Univ. Bonn, Bonner Meteorol. Abhandl.* **5**.

Garratt, J. R. (1977). Review of drag coefficients over oceans and continents. *Mon. Weather Rev.* **105**, 915–928.

Gifford, F. A., and Hanna, S. R. (1973). Modelling urban air pollution. *Atmos. Environ.* **7**, 131–136.

Goudeau, D. A., and Conner, W. C. (1968). Storm surge over the Mississippi River delta accompanying Hurricane Betsy, 1965. *Mon. Weather Rev.* **96**, 118–124.

Graf, W. H., Merz, N., and Perrinjaquet, C. (1984). Aerodynamic drag measured at a nearshore platform on Lake of Geneva. *Arch. Meteorol. Geophys. Bioklim.* **A33**, 151–173.

Haltiner, G. J., and Martin, F. L. (1957). "Dynamical and Physical Meteorology." McGraw-Hill, New York.

Hanna, S. R. (1985). Air quality modeling over short distances. *In* "Handbook of Applied Meteorology" (D. D. Houghton, ed.), pp. 712–743. Wiley, New York.

Hanson, H. P., and Long, B. (1985). Climatology of cyclogenesis over the East China Sea. *Mon. Weather Rev.* **113**, 697–707.

Hart, J. E., Rao, G. V., Van de Boogaard, H., Young, J. A., and Findlater, J. (1978). Aerial observations of the East African low-level jet stream. *Mon. Weather Rev.* **106**, 1714–1724.

Hasse, L., and Weber, H. (1985). On the conversion of Pasquill categories for use over sea. *Boundary-Layer Meteorol.* **31**, 177–185.

Hasselmann, K., Ross, D. B., Muller, P., and Sell, W. (1976). A parametric wave prediction model. *J. Phys. Oceanogr.* **6**, 200–228.

Haugen, D. A., ed. (1973). "Workshop on Micrometeorology." American Meteorological Society, Boston.

Hauwritz, B. (1947). Comments on the sea breeze circulation. *J. Meteorol.* **4**, 1–8.

Heaps, N. S. (1969). A two-dimensional numerical sea model. *Philos. Trans. R. Soc. London, Ser. A* **265**, 93–137.

Hebert, P. J. (1980). North Atlantic tropical cyclones, 1979. *Mariners Weather Log* **24**, 88–103.

Hess, S. L. (1959). "Introduction to Theoretical Meteorology." Holt, New York.

Hewson, E. W., and Olsson, L. E. (1967). Lake effects on air pollution dispersion. *J. Air Pollut. Control Assoc.* **17**, 757–761.

Hoel, P. G. (1965). "Introduction to Mathematical Statistics," 3rd Ed. Wiley, New York.

Hoffman, G. L., and Duce, R. A. (1972). Consideration of the chemical fractionation of alkali and alkaline earth metals in the Hawaiian marine atmosphere. *J. Geophys. Res.* **77**, 5161–5169.

Högström, U. (1985). Von Karman constant in atmospheric boundary layer flow: Reevaluated. *J. Atmos. Sci* **42**, 263–270.

Högström, U., and Smedman-Högstrom, A.-S. (1984). The wind regime in coastal areas with special reference to results obtained from the Swedish Wind Energy Program. *Boundary-Layer Meteorol.* **30**, 351–373.

Holladay, C. G., and O'Brien, J. J. (1975). Mesoscale variability of sea surface temperatures. *J. Phys. Oceanogr.* **5**, 761–772.

Holton, J. R. (1979). "An Introduction to Dynamic Meteorology," 2nd Ed. Academic Press, New York.

Holzworth, G. C. (1972). Mixing heights, wind speeds, and potential for urban air pollution throughout the contiguous United States. Office of Air Programs Publ. No. AP-101,

Environmental Protection Agency, Research Triangle Park, North Carolina.

Horikawa, K., Hotta, S., and Kraus, N. C. (1986). Literature review of sand transport by wind on a dry sand surface. *Coastal Eng.* **9**, 503–526.

Houze, R. A., Jr., Geotis, S. G., Marks, F. D., Jr., and West, A. I. (1981). Winter monsoon convection in the vicinity of North Borneo. Part I: Structure and time variation of the clouds and precipitation. *Mon. Weather Rev.* **109**, 1595–1614.

Hoyt, H. J. (1966). Air and sand movements to the lee of dunes. *Sedimentology* **7**, 137–143.

Hsu, S. A. (1967). Mesoscale surface temperature characteristics of the Texas Coast sea breeze. Technical Report No. 6, Atmospheric Science Group, College of Engineering, University of Texas, Austin.

Hsu, S. A. (1969). Mesoscale structure of the Texas Coast sea breeze. Technical Report No. 16, Atmospheric Science Group, College of Engineering, University of Texas, Austin.

Hsu, S. A. (1970). Coastal air-circulation system: Observations and empirical model. *Mon. Weather Rev.* **98**, 487–509.

Hsu, S. A. (1971). Wind stress criteria in eolian sand transport. *J. Geophys. Res.* **76**, 8684–8686.

Hsu, S. A., Giglioli, M. E. C., Reiter, P., and Davies, J. (1972a). Heat and water balance studies on Grand Cayman. *Caribb. J. Sci.* **12**, 9–22.

Hsu, S. A. (1972b). Boundary-layer trade-wind profile and stress on a tropical coast. *Boundary-Layer Meteorol.* **2**, 284–289.

Hsu, S. A. (1973). Dynamics of the sea breeze in the atmospheric boundary layer: A case study of the free convection regime. *Mon. Weather Rev.* **101**, 187–194.

Hsu, S. A. (1974a). A dynamic roughness equation and its application to wind stress determination at the air-sea interface. *J. Phys. Oceanogr.* **4**, 116–120.

Hsu, S. A. (1974b). On the log-linear wind profile and the relationship between shear stress and stability characteristics over the sea. *Boundary-Layer Meteorol.* **6**, 509–514.

Hsu, S. A. (1974c). Experimental results of the drag coefficient estimation for air-coast interfaces. *Boundary-Layer Meteorol.* **6**, 505–507.

Hsu, S. A. (1976). Determination of the momentum flux at the air-sea interface under variable meteorological and oceanographic conditions: Further application of the wind-wave interaction method. *Boundary-Layer Meteorol.* **10**, 221–226.

Hsu, S. A. (1977). Boundary-layer meteorological research in the coastal zone. In "Geoscience and Man" (H. J. Walker, ed.), Vol. 18, pp. 99–111. School of Geoscience, Louisiana State Univ., Baton Rouge.

Hsu, S. A. (1978). Micrometeorological fluxes in estuaries. In "Estuarine Transport Processes" (B. Kjerfve, ed.), pp. 125–134. Univ. of South Carolina Press, Columbia.

Hsu, S. A. (1979a). Mesoscale nocturnal jetlike winds within the planetary boundary layer over a flat, open coast. *Boundary-Layer Meteorol.* **17**, 485–495.

Hsu, S. A. (1979b). An operational forecasting model for the variation of mean maximum mixing height across the coastal zone. *Boundary-Layer Meteorol.* **16**, 93–98.

Hsu, S. A. (1980). Transfer of heat on a tropical beach. *Caribb. J. Sci.* **15**, 159–163.

Hsu, S. A. (1981a). Relationship between monthly frontal overrunning and offshore-onshore temperature differences across the central Gulf Coast. *J. Appl. Meteorol.* **20**, 1479–1482.

Hsu, S. A. (1981b). Models for estimating offshore winds from onshore meteorological measurements. *Boundary-Layer Meteorol.* **20**, 341–351.

Hsu, S. A. (1982a). Some mesoscale boundary-layer processes over coastal waters. *Int. Conf. Meteorol. Air/Sea Interact. Coastal Zone, 1st* pp. 298–303.

Hsu, S. A. (1982b). Determination of the power-law wind profile exponent on a tropical coast. *J. Appl. Meteorol.* **21**, 1187–1190.

Hsu, S. A. (1983a). On the growth of a thermally modified boundary layer by advection of warm air over a cooler sea. *J. Geophys. Res.* **88**, 771–774.

Hsu, S. A. (1983b). Measurements of the height of the convective surface boundary layer over an arid coast on the Red Sea. *Boundary-Layer Meteorol.* **26**, 391–396.

Hsu, S. A. (1983c). Determining latent heat flux at sea, a comparison between wind-wave interaction and profile methods. *Boundary-Layer Meteorol.* **25**, 417–421.

Hsu, S. A. (1984a). Effect of cold-air advection on internal boundary-layer development over warm oceanic currents. *Dyn. Atmos. Oceans* **8**, 307–319.

Hsu, S. A. (1984b). Sea-breeze-like winds across the north wall of the Gulf Stream: An analytical model. *J. Geophys. Res.* **89**, 2025–2028.

Hsu, S. A. (1986a). A mechanism for the increase of wind stress (drag) coefficient with wind speed over water surfaces: A parametric model. *J. Phys. Oceanogr.* **16**, 144–150.

Hsu, S. A. (1986b). Correction of land-based wind data for off-shore applications: A further evaluation. *J. Phys. Oceanogr.* **16**, 390–394, 1986.

Hsu, S. A. (1986c). A note on estimating the height of the convective internal boundary layer near shore. *Boundary-Layer Meteorol.* **35**, 311–316.

Hsu, S. A., and Whelan, T., III (1976). Transport of atmospheric sea salt in the coastal zone. *Environ. Sci. Technol.* **10**, 281–283.

Hsu, S. A., Fett, R., and La Violette, P. E. (1985). Variations in atmospheric mixing height across oceanic thermal fronts. *J. Geophys. Res.* **90**, 3211–3224.

Huschke, R. E., ed. (1959). "Glossary of Meteorology." American Meteorological Society, Boston.

Inman, D. L., Ewing, G. C., and Corliss, J. B. (1966). Coastal sand dunes of Guerrero Negro, Baja California, Mexico. *Geol. Soc. Am. Bull.* **77**, 787–802.

Ishimaru, A. (1985). Wave propagation. *In* "Handbook of Applied Meteorology" (D. D. Houghton, ed.), pp. 1031–1064. Wiley, New York.

James, N. P. (1972). Holocene and Pleistocene calcareous crust (Caliche) profiles, criteria for subaerial exposure. *J. Sediment Petrol.* **42**, 817–836.

James, R. W. (1966). Ocean thermal structure forecasting. SP 105, U.S. Naval Oceanographic Office, NSTL Station, Mississippi.

Janssen, P. A. E. M., and Komen, G. J. (1985). Effect of atmospheric stability on the growth of surface gravity waves. *Boundary-Layer Meteorol.* **32**, 85–96.

Jarvinen, B. R., and Lawrence, M. B. (1985). An evaluation of the SLOSH storm-surge model. *Bull. Am. Meteorol. Soc.* **66**, 1408–1411.

Jelesnianski, C. P. (1972). SPLASH—Special Program to List Amplitudes of Surges from Hurricanes. National Oceanic and Atmospheric Administration, Rockville, Maryland.

Johnson, G. A., Meindl, E. A., Mortimer, E. B., and Lynch, J. S. (1984). Features associated with repeated strong cyclogensis in the western Gulf of Mexico during the winter of 1982–1983. *Conf. Meteorol. Coastal Zone, 3rd* pp. 110–117.

Justus, C. G. (1985). Wind energy. *In* "Handbook of Applied Meteorology" (D. D. Houghton, ed.), pp. 915–944. Wiley, New York.

Kadib, A. A. (1965). A function of sand movement by wind. Univ. of California, Berkeley, Tech. Rep. HEL-2-12.

Keeler, C. M. (1970). Meteorology and climatology of the Arctic. *In* "SEV Arctic Environemnt Data Package, "pp. 63–76. U.S. Army Cold Regions Res. Eng. Lab., Hanover, New Hampshire.

Kitaigorodskii, S. A. and Volkov, Y. A. (1965). On the roughness parameter of the sea surface and the calculation of momentum flux in the near-water layer of the atmosphere. *Izv. Atmos. Oceanic Phys.* **1**, 566–574.

Komen, G. J., Hasselmann, S., and Hasselmann, K. (1984). On the existence of a fully developed wind-sea spectrum. *J. Phys. Oceanogr.* **13**, 1271–1285.

Kondo, J. (1975). Air-sea bulk transfer coefficients in diabatic conditions. *Boundary-Layer Meteorol.* **9**, 91–112.

Kotsch, W. J. (1983). "Weather for the Mariners." Naval Institute Press, Annapolis, Maryland.

Kraus, H., Malcher, J., and Schaller, E. (1985). A nocturnal low level jet during PUKK. *Boundary-Layer Meteorol.* **31**, 187–195.

Krishnamurti, T. N., and Wong, V. (1979). A planetary boundary-layer model for the Somali jet. *J. Atmos. Sci.* **36**, 1895–1907.

Landsberg, H. (1942). The structure of the wind over a sand-dune. *Am. Geophys. Union Trans.* Part II, pp. 237–239.

Large, W. G., and Pond, S. (1981). Open ocean momentum flux measurements in moderate to strong winds. *J. Phys. Oceanogr.* **11**, 324–336.

Large, W. G., and Pond, S. (1982). Sensible and latent heat flux measurements over the ocean. *J. Phys. Oceanogr.* **12**, 464–482.

La Violette, P. E. (1982). The Grand Banks Experiment: A satellite/aircraft/ship experiment to explore the ability of specialized radars to define ocean fronts. Report No. 49, Naval Ocean Research and Development Activity, NSTL Station, Mississippi.

Lettau, H. (1954). Graphs and illustrations of diverse atmospheric states and processes observed during the seventh test period of the Great Plains Turbulence Field Program. Occasional Rept. No. 1, Atmos. Anal. Lab., Geophys. Res. Div., Cambridge, Massachusetts.

Lettau, H. (1957). Windprofil, innere Reibung und Energieumsatz in den unteren 500 müber dem Meer. *Beitr. Phys. Atmosph.* **30**, 78–96.

Lettau, H. H. (1967). Small- to large-scale features of boundary layer structure over mountain slopes. *Symp. Mountain Meteorol., Colorado State Univ., Fort Collins, Sect. 2, Proc., Atmos. Sci.* Pap. No. 122.

Lettau, K., and Lettau, H. (1969). Bulk transport of sand by the barchans of the Pampa de la Joya in southern Peru. *Zeitschr. Geomorphol.* **13**, 182–195.

List, R. J. (1951/1984). "Smithsonian Meteorological Tables." Smithsonian Institution Press, Washington, D.C.

Liu, P. C., Schwab, D. J., and Bannett, J. R. (1984). Comparison of a two-dimensional wave prediction model with synoptic measurements in Lake Michigan. *J. Phys. Oceanogr.* **14**, 1514–1518.

Lyons, W. A. (1975). Turbulent diffusion and pollutant transport in shoreline environments. *Lect. Air Pollut. Environ. Impact Anal.* pp. 136–208.

Lyons, W. A., and Cole, H. S. (1973). Fumigation and plume trapping on the shores of Lake Michigan during stable onshore flow. *J. Appl. Meteorol.* **12**, 494–510.

McBean, G. A., ed. (1979). The planetary boundary layer. Tech. Note 165, World Meteorol. Organization, Geneva.

Mahrt, L., and Lenschow, D. H. (1976). Growth dynamics of the convectively mixed layer. *J. Atmos. Sci.* **33**, 41–51.

Mahrt, L., and Paumier, J. (1982). Cloud-top entrainment instability observed in AMTEX. *J. Atmos. Sci.* **39**, 622–634.

Mazzarella, D. A. (1985). Measurements today. *In* "Handbook of Applied Meteorology" (D. D. Houghton, ed.), pp. 283–328. Wiley, New York.

Mertins, H. O. (1976). Compendium of lecture notes in marine meteorology for Class III and Class IV personnel. Tech. Note 434, World Meteorol. Organization, Geneva.

Miller, J. E. (1946). Cyclogenesis in the Atlantic coastal region of the United States. *J. Meteorol.* **3**, 31–44.

Mistra, P. K. (1980). Dispersion from tall stacks into a shoreline environment. *Atmos. Environ.* **14**, 396–400.

Mistra, P. K., and Onlock, S. (1928). Modelling continuous fumigation of the Nanticoke Generating Station Plume. *Atmos. Environ.* **16**, 479–489.

Mizuno, T. (1982). On the similarity of the characteristics of turbulence in an unstable boundary layer. *Boundary-Layer Meteorol.* **23**, 69–83.

Monin, A. S., and Obukhov, A. M. (1954). Basic laws of turbulent mixing in the atmosphere near the ground. *Tr. Akad. Nauk. SSSR Geofiz. Inst.* No. 24(151), 163–187.

Muller, R. A. (1977). A synoptic climatology for environmental baseline analysis: New Orleans. *J. Appl. Meteorol.* **16**, 20–33.

Munn, R. E. (1966). "Descriptive Micrometeorology." Academic Press, New York.

Murray, S. P. (1975). Trajectories and speeds of wind-driven currents near the coast. *J. Phys. Oceanogr.* **5**, 347–360.

Murray, S. P., Hsu, S. A., Roberts, H. H., Owens, E. H., and Crout, R. L. (1982). Physical processes and sedimentation on a broad, shallow bank. *Estuarine, Coastal, Shelf Sci.* **14**, 135–157.

Murty, T. W., McBean, G. A., and McKee, B. (1983). Explosive cyclogenesis over the northeast Pacific Ocean. *Mon. Weather Rev.* **111**, 1131–1135.

Neumann, G., and Pierson, W. J., Jr. (1966). "Principles of Physical Oceanography." Prentice-Hall, Englewood Cliffs, New Jersey.

Ogawa, Y., and Ohara, T. (1985). The turbulent structure of the internal boundary layer near the shore. *Boundary-Layer Meteorol.* **31**, 369–384.

Oke, T. R. (1978). "Boundary Layer Climates." Methuen, London.

Overland, J. E., and Walter, B. A., Jr. (1981). Gap winds in the Strait of Juan de Fuca. *Mon. Weather Rev.* **109**, 2221–2233.

Pond, S., Phelps, G. T., Paquin, J. E., McBean, G., and Stewart, R. W. (1971). Measurements of the turbulent fluxes of momentum, moisture, and sensible heat over the ocean. *J. Atmos. Sci.* **28**, 901–917.

Palmén, E., and Newton, C. W. (1969). "Atmospheric Circulation Systems." Academic Press, New York.

Panofsky, H. A., and Dutton, J. A. (1984). "Atmospheric Turbulence." Wiley, New York.

Panofsky, H. A., and Peterson, E. L. (1972). Wind profiles and change of terrain roughness at Risø. *Q. J. R. Meteorol. Soc.* **98**, 845–854.

Paulson, C. A. (1970). The mathematical representation of wind speed and temperature profiles in the unstable atmospheric surface layer. *J. Appl. Meteorol.* **9**, 857–861.

Pielke, R. A. (1974a). A comparison of three-dimensional and two-dimensional numerical prediction of sea breezes. *J. Atmos. Sci.* **31**, 1577–1585.

Pielke, R. A. (1974b). A three-dimensional numerical model of the sea breezes over South Florida. *Mon. Weather Rev.* **102**, 115–139.

Pielke, R. A. (1984). "Mesoscale Meteorological Modeling." Academic Press, New York.

Pierson, W. J., and Moskowitz, L. (1964). A proposed spectral form for fully developed wind seas based on the similarity theory of S. A. Kitaigorodskii. *J. Geophys. Res.* **69**, 5181–5190.

Plate, E. J. (1971). Aerodynamic characteristics of atmospheric boundary layers. U.S. Atomic Energy Commission. Available as TID-25465 from National Technical Information Service, U.S. Dept. of Commerce, Springfield, Virginia 22151.

Platzman, G. W. (1963). The dynamical prediction of wind tides on Lake Erie. *Meteorol. Monogr.* No. 26.

Pond, S., Phelps, G. T., Paquin, J. E., McBean, G, A., and Stewart, R. W. (1971). Measurements of the turbulence fluxes of momentum moisture and sensible heat over the ocean. *J. Atmos. Sci.* **28**, 901–917.

Portig, W. H. (1965). Central American rainfall. *Geogr. Rev.* **55**, 68–90.

Portig, W. H. (1972). The climate of Central America. *World Surv. Climatol.* **12**, 405–478.

Powell, M. D. (1982). The transition of the Hurricane Frederic boundary-layer wind field from the open Gulf of Mexico to landfall. *Mon. Weather Rev.* **110**, 1912–1932.

Priestley, C. H. B. (1959). "Turbulent Transfer in the Lower Atmosphere." Univ. of Chicago Press, Chicago.

Quirox, R. S. (1983). The climate of the "El Nino," winter of 1982–83—a season of extraordinary climate anomalies. *Mon. Weather Rev.* **111**, 1685–1706.

Rasmussen, J. L., Murray, W. L., and Greenfield, R. S. (1976). Conference summary of the GATE research program. *Bull. Am. Meteorol. Soc.* **57**, 1005–1011.

Raynor, G. S., Michael, P., Brown, R. M., and SethuRaman, S. (1975). Studies of atmospheric diffusion from a nearshore oceanic site. *J. Appl. Meteorol.* **14**, 1080–1094.

Reed, R. J. (1980). Destructive winds caused by an orographically induced mesoscale cyclone. *Bull. Am. Meteorol. Soc.* **61**, 1346–1355.

Resio, P. T., and Vincent, C. L. (1977). Estimation of winds over the Great Lakes. *J. Waterways Harbors Coastal Div., ASCE* **102**, 263–282.

Riehl, H. (1979). "Climate and Weather in the Tropics." Academic Press, London.

Rifai, M. F., and Smith, K. V. H. (1971). Flow over triangular elements simulating dunes. *Am. Soc. Civil Eng. J. Hydraul. Div.* **93** (HY7), 963–976.

Roll, H. U. (1965). "Physics of the Marine Atmosphere." Academic Press, New York.

Rubenstein, D. M. (1981). The daytime evolution of the East African jet. *J. Atmos. Sci.* **38**, 114–128.

Saucier, W. J. (1949). Texas-West Gulf cyclones. *Mon. Weather Rev.* **77**, 219–231.

Saucier, W. J. (1955). "Principles of Meteorological Analysis." Univ. of Chicago Press, Chicago.

Schmidt, R. A. (1986). Transport rate of drifting snow and the mean wind speed profile. *Boundary-Layer Meteorol.* **34**, 213–241.

Schols, J. L. J., and Wartena, L. (1986). A dynamical description of turbulent structures in the near neutral atmospheric surface layer: The role of static pressure fluctuations. *Boundary-Layer Meteorol.* **34**, 1–15.

Schwab, D. J. (1978). Simulation and forecasting of Lake Erie storm surges. *Mon. Weather Rev.* **106**, 1476–1487.

Schwab, D. J. (1981). Determination of wind stress from water level fluctuation. Ph.D. thesis, University of Michigan.

Schwab, D. J. (1983). Numerical simulation of low-frequency current fluctuations in Lake Michigan. *J. Phys. Oceanogr.* **13**, 2213–2224.

Scorer, R. S. (1958). "Natural Aerodynamics." Pergamon, New York.

Sedefian, L. (1980). On the vertical extrapolation of mean wind power density. *J. Appl. Meteorol.* **19**, 488–493.

Sellers, W. D. (1965). "Physical Climatology." Univ. of Chicago Press, Chicago.

SethuRaman, S., and Raynor, G. S. (1980). Comparison of mean wind speeds and turbulence at a coastal site and offshore location. *J. Appl. Meteorol.* **19**, 15–21.

Sheppard, P. A., Charnock, H., and Francis, J. R. D. (1952). Observations of the Westerlies over the sea. *Q. J. R. Meteorol. Soc.* **78**, 563–582.

Simons, T. J. (1974). Verification of numerical models of Lake Ontario. I. Circulation in spring and early summer. *J. Phys. Oceanogr.* **4**, 507–523.

Simons, T. J. (1975). Verification of numerical models of Lake Ontario. II. Stratified circulation and temperature changes. *J. Phys. Oceanogr.* **5**, 98–110.

Simpson, R. H., and Riehl, H. (1981). "The Hurricane and Its Impact." Louisiana State Univ. Press, Baton Rouge.

Smedman, A.-S., and Högström, U. (1983). Turbulent characteristics of a shallow convective internal boundary layer. *Boundary-Layer Meteorol.* **25**, 271–287.

Smith, S. D. (1980). Wind stress and heat flux over the ocean in gale force winds. *J. Phys. Oceanogr.* **10**, 709–726.

Stage, S. A., and J. A. Businger (1981). A model for entrainment into a cloud-topped marine boundary layer. I. Model description and application to a cold-air outbreak episode. *J. Atmos. Sci.* **38**, 2213–2229.

Steyn, D. G., and Oke, T. R. (1982). The depth of the daytime mixed layer at two coastal sites: A model and its validation. *Boundary-Layer Meteorol.* **24**, 161–180.

Stommel, H. (1966). "The Gulf Stream, a Physical and Dynamical Description." Univ. of California Press, Berkeley.

Stull, R. B. (1976). The energetics of entrainment across a density interface. *J. Atmos. Sci.* **33**, 1260–1267.

Stunder, M., and SethuRaman, S. (1985). A comparative evaluation of the coastal internal boundary layer height equations. *Boundary-Layer Meteorol.* **32**, 177–204.

Sutton, O. G. (1953). "Micrometeorology." McGraw-Hill, New York.

Svasek, J. N., and Terwindt, J. H. J. (1974). Measurements of sand transport by wind on a natural beach. *Sedimentology* **21**, 311–322.

Sverdrup, H. U., and Munk, W. H. (1947). "Wind, Sea, and Swell: Theory of Relations for Forecasting." Publ. No. 601, U. S. Navy Hydrographic Office, Washington, D.C.

Sweet, W., Fett, R., Kerling, J., and La Violette, P. (1981). Air-sea interaction effects in the lower troposphere across the north wall of the Gulf Stream. *Mon. Weather Rev.* **109**, 1042–1052.

Tennekes, H. (1973). Similarity laws and scale relations in planetary boundary layers. *In* "Workshop on Micrometerology" (D. A. Haugen, ed.), pp. 177–216. American Meteorological Society, Boston.

Tennekes, H., and Driedonks, A. G. M. (1981). Basic entrainment equations for the atmospheric boundary layer. *Boundary-Layer Meteorol.* **20**, 515–531.

Timmerman, H. (1977). Meteorological effects on tidal heights in the North Sea. *Meded. Verh.* **99**.

Turner, D. P. (1969). "Workbook of Atmospheric Dispersion Estimates." U.S.E.P.A., Office of Air Programs, Publ. No. AP–26.

Uccellini, L. W., Kocin, P. J., Petersen, R. A., Wash, C. H., and Brill, K. F. (1984). The President's Day cyclone of 18–19 February 1979; Synoptic overview and analysis of the subtropical jet streak influencing the pre-cyclogenetic period. *Mon. Weather Rev.* **112**, 31–55.

U. S. Army Corps of Engineers (1984). "Shore Protection Manual," Vol. I. Superintendent of Documents, U. S. Government Printing Office, Washington, D.C. 20402.

U.S. National Weather Service. (1965). Project *Stormfury* Annual Report, Appendix D, 1965. National Hurricane Center, Miami, Florida.

Van Dop, H., Steenkist, R., and Nieuwstadt, F. T. M. (1979). Revised estimates for continuous shoreline fumigation. *J. Appl. Meteorol.* **18**, 133–137.

Venkatram, A. (1977). A model of internal boundary-layer development. *Boundary-Layer Meteorol.* **11**, 419–437.

Venkatram, A. (1985). Air quality modeling over long distances. *In* "Handbook of Applied Meteorology" (D. D. Houghton, ed.), pp. 744–753. Wiley, New York.

Wallace, J. M., and Hobbs, P. V. (1977). "Atmospheric Science: An Introductory Survey." Academic Press, New York.

Walter, B. A., Jr., and Overland, J. E. (1982). Response of stratified flow in the lee of the Olympic Mountains. *Mon. Weather Rev.* **110**, 1458–1473.

Walters, C. D. (1973). Atmospheric surface boundary layer wind structure studies on an Alaskan Arctic coast during winter and summer seasons. M. S. thesis, Louisiana State Univ., Baton Rouge.

Webster, P. J. (1983). Large-scale structure of the tropical atmosphere. *In* "Large-Scale Dynamical Processes in the Atmosphere" (B. J. Hoskins and R. P. Pearce, eds.), pp. 235–275. Academic Press, New York.

Wieringa, J. (1980). Representativeness of wind observation at airports. *Bull. Am. Meteorol. Soc.* **61**, 962–971.

Williams, J., Higginson, J. J., and Rohrbough, J. D. (1968). "Sea and Air." U. S. Naval Institute, Annapolis, Maryland.

Wu. J. (1982). Wind-stress coefficients over sea surface from breeze to hurricane. *J. Geophys. Res.* **87**, 9704–9706.

Wyngaard, J. C. (1973). On surface-layer turbulence. *In* "Workshop on Micrometeorology" (D. A. Haugen, ed.), pp. 101–149. American Meteorological Society, Boston.

Wyngaard, J. C., ed. (1978). "Workshop on the Planetary Boundary Layer, 14–18 August 1978, Boulder, Colorado." American Meteorological Society, Boston.

Wyngaard, J. C., and Cote, O. R. (1971). The budgets of turbulent kinetic energy and temperature variance in the atmospheric surface layer. *J. Atmos. Sci.* **28**, 190–201.

Wyngaard, J. C., Pennell, W. T., Lenschow, D. H., and LeMone, M. A. (1978). The temperature-humidity covariance budget in the convective boundary layer. *J. Atmos. Sci.* **35**, 153–164.

Yoshino, M. M. (1975). "Climate in a Small Area." Univ. of Tokyo Press, Tokyo.

Yoshino, M. M. (1976). "Local Wind 'Bora'." Univ. of Tokyo Press, Tokyo.

Index